Data-Driven Methods for Adaptive Spoken Dialogue Systems

Oliver Lemon • Olivier Pietquin
Editors

Data-Driven Methods for Adaptive Spoken Dialogue Systems

Computational Learning for Conversational Interfaces

Springer

Editors
Oliver Lemon
Mathematics and Computer Science
Heriot Watt University
Edinburgh, UK

Olivier Pietquin
SUPELEC - UMI 2958 (GeorgiaTech - CNRS)
Metz Campus - IMS - MaLIS Research Group
rue Edouard Belin 2
Metz, France

ISBN 978-1-4614-4802-0 ISBN 978-1-4614-4803-7 (eBook)
DOI 10.1007/978-1-4614-4803-7
Springer New York Heidelberg Dordrecht London

Library of Congress Control Number: 2012946737

Printed on acid-free paper

Springer is part of Springer Science+Business Media (www.springer.com)

Acknowledgements

In addition to the contributors to this book, we would like to thank all of the researchers who worked together on the CLASSiC project:

Cedric Boidin
Phillipe Bretier
Paul Crook
Herve Frezza-Buet
Matthieu Geist
Helen Hastie
Michel Ianotto
Ioannis Konstas
Nicolas Merigaud
Lonneke van der Plas
Paola Merlo
Johanna Moore
Romain Laroche
Xingkun Liu
Ghislain Putois
Kai Yu

We would also like to thank our Project Officers and Reviewers from the EC, who gave us excellent critical guidance:

Louis ten Bosch
Phillipe Gelin (Project Officer)
Xavier Gros (Project Officer)
Mike McTear
Andy Way

The research leading to these results received funding from the European Community's Seventh Framework Programme (FP7) under grant agreement no. 216594 (CLASSiC project).

Contents

Contributors

Senthilkumar Chandramohan SUPELEC, Metz, France

Milica Gašić Cambridge University, Cambridge, UK

Helen Hastie Heriot-Watt University, Edinburgh, UK

James Henderson Université de Genève, Geneva, Switzerland

Srini Janarthanam Heriot-Watt University, Edinburgh, UK

Filip Jurčíček Charles University, Prague, Czech Republic

Simon Keizer Heriot-Watt University, Edinburgh, UK

Oliver Lemon Mathematics and Computer Science, Heriot-Watt University, Edinburgh, UK

Roberto Pieraccini SpeechCycle, New York, NY, USA

ICSI, Berkeley, CA, USA

Olivier Pietquin SUPELEC, Supélec Campus de Metz, France

Verena Rieser Heriot-Watt University, Edinburgh, UK

Stéphane Rossignol SUPELEC, Metz, France

David Suendermann SpeechCycle, New York, NY, USA

DHBW, Stuttgart, Germany

Blaise Thomson Cambridge University, Cambridge, UK

Steve Young Cambridge University, Cambridge, UK

Chapter 1
Conversational Interfaces

Oliver Lemon

Although long anticipated by science fiction authors, in 2012 speech interfaces have now arrived in everyday use. Late in 2011, Apple introduced "Siri", a speech-enabled personal assistant for smartphones. In addition, Android phones have employed speech-activated "Voice Actions" before the arrival of Siri, and in 2012 Google is rumoured to be developing a Siri-style interface with its "Majel" or "Assistant" project. Many other speech-enabled applications have also been deployed (e.g. Evi, Vlingo), and the market for speech applications is growing rapidly. Likewise, Microsoft's "Kinect" controller has added new speech input capabilities to video game controllers.

In consequence, these are exciting times for speech researchers, as the potential of spoken natural language interaction is now being understood outside the research community. However, it is important to note that Siri and its competitors are only beginning to explore the possibilities of truly *conversational* interaction. Such systems have quite limited language understanding capabilities and employ a limited model (if any) of conversational context. The usefulness of Siri perhaps derives as much from its ability to combine multiple information sources (calendar, address book, web search, Wolfram Alpha, etc) as from its current natural language processing capabilities.

To elaborate, one of the main shortcomings of current commercial voice interfaces is in their limited ability to understand and manage *extended* conversational interactions, that is interactions which span multiple turns of a dialogue. Looking

O. Lemon (✉)
Mathematics and Computer Science, Heriot-Watt University, Edinburgh, EH14 4AS, UK
e-mail: o.lemon@hw.ac.uk

O. Lemon and O. Pietquin (eds.), *Data-Driven Methods for Adaptive Spoken Dialogue Systems*, DOI 10.1007/978-1-4614-4803-7_1,

at the current examples of interaction with Siri[1] and Kinect[2], we can see that they are not actually *conversations* (i.e. with meaning being composed, edited, and responded to over several turns), but the interactions are normally single-turn inputs which are dealt with independently. (Exceptions to this are a few cases such as the system allowing revision of the previous turn, such as the following: "What's the weather in Palo Alto today?" [Siri displays a weather website] followed by "How about Sunnyvale?")

In addition, the information presented to the user by current voice search systems is still often in the form of a webpage with a list of search results, rather than being a natural language summary tailored to a particular user and their situation. This means that true extended *conversational* interaction (rather than simply *spoken* interaction) with commercial devices and services is still some way off in the future.

The CLASSiC project[3] was a European initiative working on a fully data-driven architecture for the development of conversational interfaces, as well as on new machine learning approaches for their sub-components. CLASSiC was active at around the same time as the development of Google's Voice Actions, Apple's Siri, and Microsoft's Kinect. It developed a variety of novel statistical methods for spoken dialogue processing, specifically for extended conversational interaction, which are now collected together in this book. A major focus of the project was in tracking the accumulation of information about user goals over multiple dialogue turns (i.e. extended conversational interaction) and in maintaining overall system robustness even when speech recognition results contain errors. Other advances were made in the areas of adaptive natural language generation (NLG), statistical methods for spoken language understanding (SLU), and machine learning methods for system optimisation, either during online operation or from small amounts of data. Industrial applications of these techniques were also explored in the project.

The project was successful in its lifetime in producing many peer-reviewed scientific publications as well as a number of demonstration systems illustrating the new models that it developed. At the conclusion of the project, it was decided to create this book, which collects together the main research results and lessons learned in the CLASSiC project. Each chapter herein provides a retrospective summary of the specific methods developed and results obtained in its particular research area. In addition, we also invited leading researchers in statistical methods applied to industrial-scale dialogue systems (from SpeechCycle) to contribute a chapter surveying their recent work. The book also contains an outlook for future research issues in applying machine learning methods to the challenges of SDS.

We therefore hope that this book will serve as a valuable introduction to the current state of the art in statistical approaches to developing conversational

[1] See http://i.tuaw.com/2011/10/05/iphone-4s-what-can-you-say-to-siri/ and http://www.apple.com/iphone/features/siri.html.

[2] See http://support.xbox.com/en-US/kinect/voice/speech-recognition.

[3] "Computational Learning in Adaptive Systems for Spoken Conversation", European Community FP7, 2008–2011, project number 216594, www.classic-project.org.

interfaces, for active researchers in the field in industry and academia, as well as for students who are considering working in this exciting area.

In recent developments, and following on from the work of the CLASSiC project, the PARLANCE project[4] is now developing these methods further, for incremental "hyperlocal" conversational interaction in several languages (see Chap. 9 for details).

1.1 Structure of the Book

Following this introduction, Chap. 2 outlines some of the basic concepts and problems for statistical approaches to the optimisation of SDS, including an introduction to the use of Markov decision processes, and a presentation of a data-driven methodology using "Wizard-of-Oz" data collections, which allows development of dialogue managers from limited amounts of data. It also surveys a variety of datasets which are now available for developers and researchers in statistical approaches to SDS, in particular, the new datasets made available by the CLASSiC project.

In Chap. 3 the overall problems of SLU are first discussed, and recent advances in data-driven SLU systems are described in general. Then, three SLU systems developed in the CLASSiC project are presented, with special attention to the issues of robustness to ill-formed user input, coping with small training sets, and evaluation of SLU components.

Chapter 4 provides an overview of the user simulations that are often employed for training and testing dialogue policies, together with a discussion of metrics and methods for evaluating such simulated users. Three techniques are covered in detail: agenda-based simulation, dynamic Bayesian network-based simulation, and inverse reinforcement learning.

Chapter 5 describes the framework of partially observable Markov decision processes, and focuses on efficient algorithms for optimising dialogue policies. Three approaches to optimisation are described, which allow faster online learning than previous work.

Chapter 6 moves on to explain current work in statistical approaches to decision-making for NLG components. Several aspects of trainable NLG systems are surveyed, including optimisation of high-level information presentation decisions (e.g. content structuring), online adaptation of referring expressions for different types of users, and the use of crowd-sourced data to train a statistical surface realiser. The evaluation results presented for these approaches show that by optimising NLG components significant benefits can be gained for overall SDS performance, including user's understanding of the information conveyed by speech.

---

[4]Funded by the European Community under FP7 2011–2014, project number 113806, www.parlance-project.eu.

Evaluation methods and metrics for spoken dialogue systems are the focus of Chap. 7. This chapter covers both objective and subjective evaluation techniques and discusses a variety of recent proposals for usability analysis. It also discusses how evaluation data and analysis can be used in informing the objective functions used by machine learning techniques. Perspectives from both academic research and industrial system development are covered, and an emerging focus on system evaluation "in the wild" is discussed.

In Chap. 8, some recent data-driven approaches applied in industrial spoken dialogue systems are presented. These range from the benefits of using statistical language models for speech recognition to optimisation of different dialogue management decisions, such as whether to transfer a call to a human operator or what order to ask questions in. This chapter shows the real benefits that are now being obtained by deployment of statistical methods in commercially deployed SDS.

We conclude the book in Chap. 9 with a summary of the main results presented, and a discussion of the current research challenges for this field in 2012 and beyond.

Chapter 2
Developing Dialogue Managers from Limited Amounts of Data

Verena Rieser and Oliver Lemon

2.1 Introduction

One of the central problems in developing a spoken dialogue system (SDS) is in how the system makes the decision of "what to say next" at any specific point in a conversation. This selection of an appropriate action is the core problem of dialogue management (DM), and it depends on having a representation of the conversational context at each decision point. This context information could consist of, for example, what information has already been conveyed in the dialogue, what the user has said in the preceding utterance (according to a speech recogniser), and the length of the dialogue so far. Making decisions regarding what to say next has been approached in a variety of ways.

Most often, these decisions have been made using a variety of rule-based approaches, sometimes consisting of many interacting rules, which together determine "what to say next" depending on different features of the current context. For example, previous research systems, such as TrindiKit [21] and DIPPER [5], implemented these rules using the "Information State Update" approach to dialogue management, whereas in industry, standards such as Voice XML[1] are common practice [27]. The required rule-sets in such approaches need to be designed and optimised by human experts, and they are difficult to maintain.

On the other hand, statistical machine learning approaches, such as reinforcement learning (RL), for SDSs offer several potential advantages over rule-based hand-crafted approaches to dialogue systems development: a data-driven development

---

[1] See www.voicexml.org.

V. Rieser (✉) • O. Lemon
Heriot-Watt University, Edinburgh, EH14 4AS, UK
e-mail: v.t.rieser@hw.ac.uk; o.lemon@hw.ac.uk

O. Lemon and O. Pietquin (eds.), *Data-Driven Methods for Adaptive Spoken Dialogue Systems*, DOI 10.1007/978-1-4614-4803-7_2,

cycle, provably optimal action policies, a precise mathematical model for action selection, the ability to generalise to unseen states, and automatic optimisation of policy trade-offs via an objective function [22, 51].

One of the major strengths of strategies developed using such methods is that they can "intelligently" adapt their action selection to the changing dialogue environment (i.e. variables representing dialogue context features) in order to satisfy an overall long-term objective (e.g. to satisfy the user's information needs). One of the major challenges for such approaches is that they rely on the availability of quantities of appropriate data for training. In this chapter, we introduce the general methods used in RL for dialogue policy learning, and we discuss methods for overcoming data limitations associated with machine learning for dialogue policies. Section 2.2 provides a brief introduction to RL-based policy learning, including the use of feature-based state representations and function approximation methods. Section 2.3 discusses requirements that data sets need to meet in order to be suitable for learning RL-based strategies and also surveys several currently available data sets, some of which are newly available in the CLASSiC project data archive. We argue that suitable data sets are scarce and, in most cases, new in-domain data sets have to be collected from scratch. To address this challenge, in Sect. 2.4, we present a framework for "bootstrapping" RL-based strategies from small data sets collected using a "Wizard-of-Oz" (WOZ) methodology. In Sect. 2.5, we briefly summarise related work which also tackles the challenge of learning from limited data (see also Chap. 5).

2.2 Background: Reinforcement Learning for SDS Policy Optimisation

The basic model for each of the approaches that we will discuss below is the Markov decision process or MDP (see [44] for an introduction). Here, a stochastic system interacting with its environment (in our case, the user of the SDS) through its actions is described by a number of states $s_i \in S$ in which a given number of actions $a_j \in A$ can be performed. In a spoken dialogue system, the states represent possible dialogue contexts, for example, the user's information goal and the dialogue history,[2] and the actions correspond to the choices in system behaviour that one wants to optimise.

Each state-action pair is associated with a transition probability $\mathcal{T}_{ss'}^{a}$: the probability of moving from state s at time t to state s' at time $t+1$ after having performed action a when in state s. For the case of SDSs, these probabilities depend on how users respond to the system's actions, as well as properties of the input

[2]Note that a common misunderstanding is that the Markov property constrains the state to exclude the dialogue history. However, we can employ variables in the current state which explicitly represent features of the history.

channel (most importantly, noise). The transitions are also associated with a reinforcement signal (or "reward") $\mathcal{R}^a_{ss'}$ describing how good the result of action a was when performed in state s. In interactive systems, these reward signals are most often associated with a final measure of task completion and the overall length/duration of the interaction. In most research systems, the reward function is defined by the system designer to optimise some a priori objectives of the system. For example, a designer might assign 100 reward points for successful task completion and -1 point per dialogue turn. However, recent work has developed data-driven methods for defining reward functions [36, 47], attempting to maximise overall user satisfaction which is discovered by analysis of data on user ratings of system behaviours.

To control a system described in this way, one needs a strategy or policy π mapping all states to actions: $\pi(s) : S \to A$. In this framework, a reinforcement learning agent is a system aiming at optimally mapping states to actions, i.e. finding the best strategy π^* so as to maximise an overall reward R which is a function (most often a weighted sum) of all the immediate rewards. In dialogue interaction, the full reward for an action is often not immediate but is *delayed* until successful completion of a task. Of course, actions affect not only the immediate reward but also the next state and thereby all subsequent rewards.

In general, then, we are trying to find an action policy π which maximises the value $Q^\pi(s,a)$ of choosing action a in state s, which is given by the Bellman equation [44]:

$$Q^\pi(s,a) = \sum_{s'} \mathcal{T}^a_{ss'} [\mathcal{R}^a_{ss'} + \gamma V^\pi(s')]. \tag{2.1}$$

(Here, we denote the expected immediate reward by $\mathcal{R}^a_{ss'}$, where γ is a discount factor between 0 and 1, and $V^\pi(s)$ is the value of state s' according to π.)

If the transition probabilities are known, an analytic solution can be computed by dynamic programming. Otherwise, the system must learn the optimal strategy using a trial-and-error process, for example, using RL algorithms such as temporal difference learning or Q-learning [44]. Trial-and-error search and delayed rewards are the two main features of RL.

2.2.1 Feature-Based State Representations

It is common practice to represent the system's states (i.e. the "dialogue context" as discussed above) using feature vectors. For example, we could have a binary feature representing whether the user has yet been greeted (1) or not (0), and a feature for each information "slot" (e.g. food type, location, price range) indicating whether it has been obtained from the user (1) or not (0). Using such a simple representation, for example, in a restaurant-search system, the state vector $\langle 1,0,0,1 \rangle$ could represent that the user has been greeted and we have obtained information on their preferred price range, but the user has not provided any information on preferred food type nor location yet. This system would have only 2^4 states and would obviously not be very expressive in distinguishing many dialogue contexts, thereby restricting the level of adaptivity possible for the system. However, much more complex feature-based representations have been used successfully. For example [16], use 1,282 features (many of which relate to the dialogue history), in a system with 74 possible actions.

2.2.2 Generalisation Methods

A benefit of feature-based state representations is that we can consider states to be similar insofar as they share features. This enables us to retrieve actions for states which we have never encountered before, by retrieving actions which are known to be good for "similar" states. This provides the important property of "generalisation," meaning that a trained policy can be used for previously unseen states.

A useful generalisation technique for RL with feature-based representations of states is linear function approximation [16]. This is used to map from a vector of features $f(s)$ for a state s to a vector of estimates $Q(s,a)$, with one estimate for each action a. The trained parameters of the linear function are a vector of weights w_a for each action. With weights trained on a data set, an estimate $Q(s,a)$ of the expected future reward given state s and action a can be made by computing the inner product of the state vector $f(s)$ and the weight vector w_a. During training, updating the estimate $Q(s,a)$ for one observed state s will also update the estimate $Q(s',a)$ for all other states s' to the extent that s' shares features with s. Each feature therefore represents a dimension with respect to which two states can be similar[3].

With these initial concepts in place, we now can discuss data requirements for learning RL-based dialogue policies.

2.3 Data Sets

2.3.1 Data Requirements for Learning RL-Based Policies

To be successful, RL-based trial-and-error learning needs to see as many example interactions as possible. Therefore, data sets used for RL-based policy optimisation should ideally explore all possible sequences of combinations of states and actions, in order to experience the different rewards available. A data set needs to include a sufficient and realistic amount of *variation* to allow this exploration, i.e. the system needs to vary the sequences of actions taken in each state.

[3]This similarity measure is known as a linear kernel, see the discussion in [16].

In addition, a suitable data set needs to contain:

- Annotations of the relevant context features for dialogue state (S) estimation, as discussed above
- Annotations of the relevant system actions (A) for optimisation
- Subjective and objective performance measures to calculate the reward function ($\mathcal{R}$), such as responses to user questionnaires, dialogue length, number of confirmations, Word-Error Rates

Learning automated dialogue strategies from human–human interaction data (if available) is not really feasible, since humans behave differently with machines than with other people [8, 19, 24]. Human-human interaction has different properties to human–machine interaction, mainly because it is (usually) less affected by channel noise since humans are much better at handling noise and uncertainty.

In sum, the ideal case is to have available one consistent, sufficiently large data set of human–machine interactions, in the domain of the target system, for learning RL-based dialogue systems. Of course, there is no fixed threshold or standard recommendation for how much data is enough. Clearly this scales with the size of the state space and the number of actions that need to be explored. Results from [1] indicate that user simulations learned from a small data set (130 dialogues) can fail to provide enough coverage in the strategy training phase and, in fact, that random models might even provide superior performance in such cases. On the other hand, results presented in Chap. 6 indicate that useful policies can be trained with simulated environments generated from as little as 12 conversations, if each conversation contains multiple observations of the phenomena of interest. Thus, the required size of data set depends on the complexity of the learning problem, as well as the chosen approach and required accuracy of the user simulations–a problem which is also described as the "triple trade-off problem" [7]. Measures for evaluating the accuracy of simulations are discussed in Chap. 4.

2.3.2 Existing Data Sets

Currently, the COMMUNICATOR corpus [3] and the CMU *Let's Go* corpus [26, 32] are two of the largest human–machine dialogue corpora available. The COMMUNICATOR corpus consists of approximately 2,300 human–machine dialogues in total. The CMU *Let's Go* bus information system has been running since March 2005 and has served over 150K calls. Henderson et al. [16] have explored the COMMUNICATOR corpus for learning an optimal dialogue policy using a combination of supervised and reinforcement learning, while [4] has used the *Let's Go* corpus for learning error recovery strategies using supervised learning.

Some other corpora have recently been collected for natural language generation (NLG) for SDSs: The GIVE corpus [12] consists of 108 (45 German and 63 American English) written interactions for instruction giving in virtual environments.

Table 2.1 Corpora collected and released by the CLASSiC project

	EuroParl (English and French)	TownInfo (English)	CamInfo (English)	Appoint-ment scheduling (French)	Appoint-ment scheduling (English)	SelfHelp (English)
DM	S.2.1	S2.2, S2.4	S2.3	S2.8, S2.9		
SLU	S2.1					
NLG		S2.10	S2.3	S2.5	S2.6	S2.7

The MATCH project also made two data sets available, which consist of 1,024 [42] and 1,224 [48] ranked utterances. Both corpora have been explored for RL-based NLG strategies by [6, 38], respectively.

2.3.3 New Data Sets Collected by the CLASSiC Project

The following corpora for training SDS were collected within the CLASSiC project. All data is available from the project data repository.[4] Further details on the experimental setup, annotation, and use of the data are described in [29]. The corpora mainly cover two domains: tourist information domain on restaurants (*TownInfo*) and bars, and hotels in Cambridge (*CamInfo*), and the appointment scheduling (AS) domain. The data has been used to train statistical approaches for spoken language understanding (SLU) [15], dialogue management policies [13,45], and NLG [17, 18, 23, 35]. The list below gives an overview of the data available (as named in the online repository: S2.1–S2.10), together with Table 2.1.

S2.1 The EuroParl parallel corpus [20] extended with 1,000 manually annotated French sentences and automatic syntactic-semantic analysis [30].
S2.2 TownInfo evaluation dialogues with the HIS system [14] (total size: 720 dialogues).
S2.3 CamInfo evaluation dialogues (total size: 709 training and 1,338 evaluation dialogues).
S2.4 TownInfo DIPPER system [5] testing three versions of a real SDS (Total size: 275 dialogues).
S2.5 French appointment-scheduling (CLASSiC system 2) data (Total size: 1.1 GB).
S2.6 NLG temporal expressions (total size: 20 MB): Log files of a web-based experiment that was conducted to gather user feedback and understanding of a variety of Temporal Expressions.
S2.7 SelfHelp WoZ dialogues (total size: 18 training and 36 evaluation dialogues): dialogue between users and WOZ system on the topic of setting-up a broadband Internet connection.

[4] http://www.macs.hw.ac.uk/iLabArchive/CLASSiCProject/Data/myaccount.php.

S2.8 Appointment scheduling (CLASSiC systems 3 and 4) (total size: 1.2 GB).
S2.9 The 1,013+ Appointment-scheduling corpus (total size: 10,000 dialogues).
S2.10 TownInfo NLG WOZ data (total size: 213 dialogues): A WOZ corpus to study information presentation strategies for SDS.

2.3.4 Discussion

When learning policies for domains which are not covered by existing data sets, new in-domain data has to be collected. However, data is expensive to collect and typically in-domain systems which show the desired variation and annotations hardly ever exist. Collecting dialogue data without a working prototype is problematic, leaving the developer with a classic "chicken-and-egg" problem.

In the following, we present an approach which enables strategy learning in domains where no prior system is available. Optimised learned strategies are then available from the first moment of online operation, and handcrafting of dialogue strategies is avoided. This independence from large amounts of in-domain dialogue data allows researchers to apply RL to new application areas beyond the scope of existing dialogue systems. We call this method "bootstrapping" [39].

2.4 A Bootstrapping Framework for Simulation-Based Learning from Small Data Sets

We now present a methodology for learning dialogue strategies using simulation-based RL, where the simulated environment is learned from small amounts of WOZ data. Recently, other researchers have also followed the framework described below, for example [10,46].

Simulation-based reinforcement learning generates learning episodes while interacting with a simulated environment, which typically consist of a user simulation [41], a noise model, e.g. [43], and any other components the policy needs to be adaptive to, such as output from a stochastic sentence generation model [35]. Chapter 4 discusses user simulations in more detail. Simulated learning environments are often used to explore optimal policies which were previously unseen in the data, for example, [1,9,50].

However, several aspects of the components of simulated environments are often hand-crafted, and thus limit the scope of policy learning. In order to build simulation components from real data, annotated in-domain dialogue corpora have to be available, meeting the requirements described in Sect. 2.3.1.

In the following, we propose to collect in-domain data using the WOZ method. In a WOZ experiment, a hidden human operator, the so-called "wizard," simulates (partly or completely) the behaviour of the application, while subjects are lead to believe that they are interacting with a real system [11]. WOZ data allows exploration of a range of possible strategies, as intuitively generated by the wizards, in contrast to using an initial system which can only explore a pre-defined range of options. However, WOZ experiments usually only produce a limited amount of data, and the optimal policy is not likely to be present in the original small data set. Our method shows how to use this data to build a simulated environment in which optimal policies can be discovered.

The use of WOZ data has earlier been proposed in the context of RL. Williams and Young [49] use WOZ data to discover the state and action space for the design of a MDP. Thomas et al. [31] use WOZ data to build a simulated user and noise model for simulation-based RL. While both studies show promising first results, their simulated environments still contain many handcrafted aspects, which makes it difficult to evaluate whether the success of the learned strategy indeed originates from the WOZ data. Schatzmann et al. [40] propose to "bootstrap" with a simulated user which is entirely hand-crafted.

For "bootstrapping" from small data sets, we use a five-step procedure, see Fig. 2.1:

1. We start by collecting data in a WOZ experiment, for example as described in [34].
2. From this data, we train and test different components of our simulated environment using supervised learning techniques. In particular, we extract a supervised policy, reflecting human (wizard) performance on this task. We build a noise simulation and two different user simulations, as well as a data-driven reward function.
3. We then train and evaluate dialogue policies by interacting with the simulated environment.
4. Once the learned policies are "good enough" in simulation, we test them with real users.
5. In addition, we introduce a final phase where we meta-evaluate the whole framework. This final step is necessary since WOZ experiments only *simulate* human–computer interaction. We therefore need to show that a strategy bootstrapped from WOZ data indeed transfers to real human–computer interaction. We first show that the results between simulated and real interaction are compatible. We also meta-evaluate the reward function, showing that it is a stable, accurate estimate for real user's preferences.

This framework was initially developed within the TALK project[5] by [33] and is further described, with a full presentation of results, in [37, 39]. It was first applied to multimodal dialogue management (where a combination of graphical display and speech is used) using a multimodal WOZ corpus of 70 dialogues with 1,772 turns [34]. The CLASSiC project also applied this method to adaptive NLG, as further described in Chap. 6.

[5] http://www.talk-project.eurice.eu/.

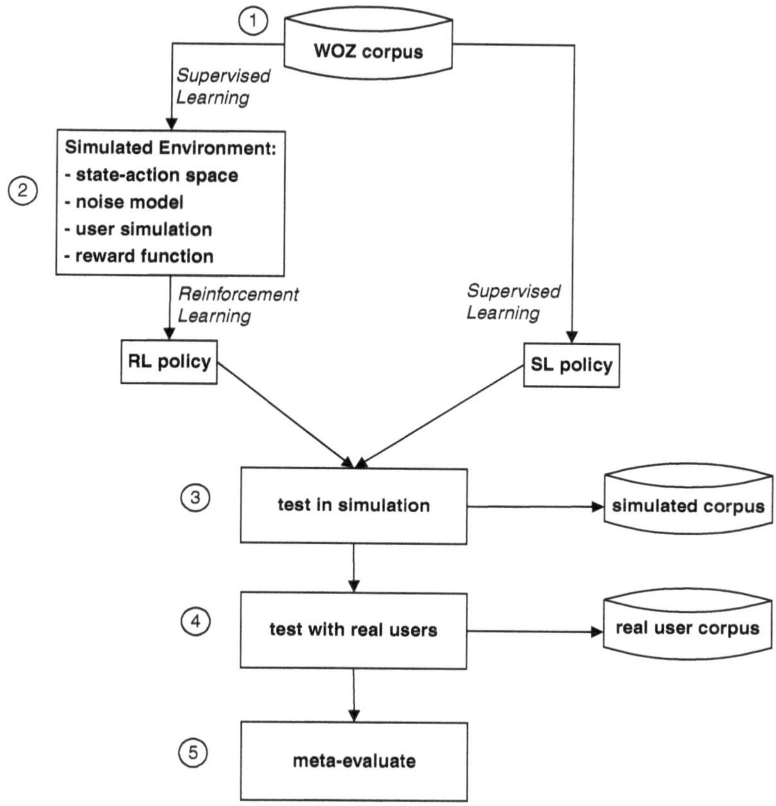

Fig. 2.1 Data-driven methodology for simulation-based dialogue strategy learning for new applications

2.5 Related Work: Methods for Learning from Small Data Sets

Simulation-based RL is a commonly applied method to deal with limited data for learning a dialogue policy. However, other methods have recently been explored within the CLASSiC project. These sample-efficient approaches allow learning of dialogue policies directly from data (and can therefore help to avoid approximations made by simulated environments).

For example, [13] use Gaussian Processes for partitioning large state spaces, which enables the RL algorithm to update state spaces more effectively and thus speeds up learning and requires less training data. Pietquin et al. [28] use approximate dynamic programming algorithms for policy optimization in SDS. These algorithms are particularly sample efficient and can learn from a few hundred dialogue examples. Please see Chap. 5 for a presentation of such techniques.

2.6 Conclusion

In this chapter, we have introduced statistical methods for optimisation of dialogue management, in particular reinforcement learning for optimising dialogue policies. We discussed data requirements for such approaches and we surveyed several existing data sets, including new data available in the CLASSiC project repository. We argued that for most cases when learning RL-based dialogue policies, in-domain data sets need to be collected from scratch. We then outlined a framework for simulation-based learning from small, "WOZ" data sets. This framework enables strategy learning in domains where neither prior data nor a working system is available. Optimised learned strategies are then available from the first moment of online operation, while avoiding handcrafting of dialogue strategies. This independence from large amounts of in-domain dialogue data allows researchers to apply RL to new application areas beyond the scope of existing dialogue systems. We call this method "bootstrapping". We also briefly discussed alternative methods for efficient policy learning, which were explored within the CLASSiC project (see Chap. 5). Note also that Chap. 6 presents two case studies applying this framework to optimising decision-making for NLG.

In future work, crowd-sourcing data collection might also be explored for policy learning, for example, using online computer games [25] or also using online multiplayer games [2].

Acknowledgements The research leading to these results has received partial support from the European Community's Seventh Framework Programme (FP7) under grant agreement no. 216594 (CLASSIC project), from the EPSRC, project no. EP/G069840/1, and from the European Community's Seventh Framework Programme (FP7) under grant agreement no. 269427 (STAC project), under grant agreement no. 270019 (SPACEBOOK project), under grant agreement no. 270435 (JAMES project), and under grant agreement no. 287615 (PARLANCE project).

References

1. Ai, H., Tetreault, J., Litman, D.: Comparing user simulation models for dialog strategy learning. In: Proc. of the North American Meeting of the Association of Computational Linguistics (NAACL), pp. 1–4. Rochester, New York, USA, April 2007
2. Asher, N., Lascarides, A., Lemon, O., Guhe, M., Rieser, V., Muller, P., Afantenos, S., Benamara, F., Vieu, L., Denis, P., Paul, S., Keizer, S., Degremont, C.: Modelling Strategic Conversation: the STAC project. The 16th workshop on the semantics and Pragmatics of Dialogue (SeineDial'12). Paris, 2012
3. Bennett, C., Rudnicky, A.: The carnegie mellon communicator corpus. In: Proc. of the International Conference of Spoken Language Processing (ICSLP), 2002
4. Bohus, D., Langner, B., Raux, A., Black, A.W., Eskenazi, M., Rudnicky, A.L: Online supervised learning of non-understanding recovery policies. In: Proc. of the IEEE/ACL workshop on Spoken Language Technology (SLT), pp. 170–173. Aruba, December 2006
5. Bos, J., Klein, E., Lemon, O., Oka, T.: DIPPER: Description and formalisation of an Information-State Update dialogue system architecture. In: Proc. of the 4th SIGdial Workshop on Discourse and Dialogue, 2002

6. Dethlefs, N., Cuayáhuitl, H.: Hierarchical reinforcement learning and hidden markov models for task-oriented natural language generation. In: Proc. of 49th Annual Meeting of the Association for Computational Linguistics, 2011
7. Dietterrich, T.G.: Machine learning. Nature Encyclopedia of Cognitive Science, 2003
8. Doran, C., Aberdeen, J., Damianos, L., Hirschman, L.: Comparing several aspects of Human-Computer and Human-Human dialogues. In: Proc. of the 2nd SIGDIAL Workshop on Discourse and Dialogue, 2001
9. Eckert, W., Levin, E., Pieraccini, R.: User modeling for spoken dialogue system evaluation. In: Proc. of the IEEE workshop on Automatic Speech Recognition and Understanding (ASRU), pp. 80–87. Santa Barbara, CA , USA, December 1997
10. Forbes-Riley, K., Litman, D.: Designing and evaluating a wizarded uncertainty-adaptive spoken dialogue tutoring system. Computer Speech and Language **25**(1), 105–126 (2011)
11. Fraser, N.M., Gilbert, G.N.: Simulating speech systems. Computer Speech and Language **5**, 81–99 (1991)
12. Gargett, A., Garoufi, K., Koller, A., Striegnitz, K.: The GIVE-2 corpus of giving instructions in virtual environments. In: Proceedings of the 7th Conference on International Language Resources and Evaluation (LREC), 2010
13. Gasic, M., Jurcicek, F., Keizer, S., Mairesse, F., Thomson, B., Yu, K., Young, S.: Gaussian processes for fast policy optimisation of a pomdp dialogue manager for a real-world task. In: Proceedings of SIGDIAL, 2010
14. Gasic, M., Keizer, S., Mairesse, F., Schatzmann, J., Thomson, B., Young, S.: Training and Evaluation of the HIS POMDP Dialogue System in Noise. In: Proc. of SIGdial Workshop on Discourse and Dialogue, 2008
15. Gesmundo, A., Henderson, J., Merlo, P., Titov, I.: A latent variable model of synchronous syntactic-semantic parsing for multiple languages. In: CoNLL 2009 Shared Task, Conf. on Computational Natural Language Learning, 2009
16. Henderson, J., Lemon, O., Georgila, K.: Hybrid Reinforcement / Supervised Learning of Dialogue Policies from Fixed Datasets. Computational Linguistics **34**(4), 487–513 (2008)
17. Janarthanam, S., Lemon, O.: Learning to Adapt to Unknown Users: Referring Expression Generation in Spoken Dialogue Systems. In: Proceedings of the 48th Annual Meeting of the Association for Computational Linguistics (ACL), pp. 69–78. Uppsala, Sweden, July 2010
18. Janarthanam, S., Hastie, H., Lemon, O., Liu, X.: 'The day after the day after tomorrow? ' A machine learning approach to adaptive temporal expression generation: training and evaluation with real users. In: Proceedings of SIGDIAL, 2011
19. Jönsson, A., Dahlbäck, N.: Talking to a computer is not like talking to your best friend. In: Proc. of the Scandinavian Conference on Artificial Intelligence, 1988
20. Koehn, P.: Europarl: A parallel corpus for statistical machine translation. In: Proceedings of the MT Summit, 2005
21. Larsson, S., Traum, D.: Information state and dialogue management in the TRINDI dialogue move engine toolkit. Natural Language Engineering Special Issue on Best Practice in Spoken Language Dialogue Systems Engineering **6**(3-4), 323–340 (2000)
22. Lemon, O., Pietquin, O.: Machine Learning for spoken dialogue systems. In: Proc. of the International Conference of Spoken Language Processing (Interspeech/ICSLP), pp. 2685–2688. Antwerp, Belgium, September 2007
23. Mairesse, F., Gašić, M., Jurčíček, F., Keizer, S., Thomson, B., Yu, K., Young, S.: Phrase-based statistical language generation using graphical models and active learning. In: Proceedings of the 48th Annual Meeting of the Association for Computational Linguistics, ACL '10, pp. 1552–1561. Stroudsburg, PA, USA (2010) Association for Computational Linguistics
24. Moore, R.K., Morris, A.: Experiences collecting genuine spoken enquiries using woz techniques. In: Proc. 5th DARPA workshop on Speech and Natural Language, 1992
25. Orkin, J., Roy, D.: The restaurant game: Learning social behavior and language from thousands of players online. Journal of Game Development **3**(1), 39–60 (2007)

26. Parent, G., Eskenazi, M.: Toward better crowdsourced transcription: Transcription of a year of the let's go bus information system data. In: Proc. of the IEEE/ACL Spoken Language Technology (SLT), 2010
27. Pieraccini, R., Suendermann, D., Dayanidhi, K., Liscombe, J.: Are we there yet? research in commercial spoken dialog systems. In: Proceedings of TSD'09, pp. 3–13, 2009
28. Pietquin, O., Geist, M., Chandramohan, S., Frezza-Buet, H.: Sample-Efficient Batch Reinforcement Learning for Dialogue Management Optimization. ACM Transactions on Speech and Language Processing **7**(3), 21 (2011)
29. Pietquin, O., Hastie, H., Janarthanam, S., Keizer, S., Putois, G., van der Plas, L.: D6.5 annotated data archive. Technical report, CLASSiC project deliverable, 2011
30. van der Plas, L., Merlo, P., Henderson, J.: Scaling up cross-lingual semantic annotation transfer. In: In Proceedings of ACL/HLT, 2011
31. Prommer, T., Holzapfel, H., Waibel, A.: Rapid simulation-driven Reinforcement Learning of multimodal dialog strategies in human-robot interaction. In: Proc. of the International Conference of Spoken Language Processing (Interspeech/ICSLP), pp. 1918–1924. Pittsburgh, Pennsylvania, USA, September 2006
32. Raux, A., Bohus, D., Langner, B., Black, A.W., Eskenazi, M.: Doing research on a deployed spoken dialogue system: One year of let's go! experience. In: Proc. of Interspeech, 2006
33. Rieser, V.: Bootstrapping Reinforcement Learning-based Dialogue Strategies from Wizard-of-Oz data. PhD thesis, Saarbrueken Dissertations in Computational Linguistics and Language Technology, vol. 28, 2008
34. Rieser, V., Kruijff-Korbayová, I., Lemon, O.: A corpus collection and annotation framework for learning multimodal clarification strategies. In: Proc. of the 6th SIGdial Workshop on Discourse and Dialogue, pp. 97–106. Lisbon, Portugal, September 2005
35. Rieser, V., Lemon, O., Liu, X.: Optimising Information Presentation for Spoken Dialogue Systems. In: Proceedings of the Annual Meeting of the Association for Computational Linguistics (ACL), pp. 1009–1018. Uppsala, Sweden, July 2010
36. Rieser, V., Lemon, O.: Learning effective multimodal dialogue strategies from Wizard-of-Oz data: Bootstrapping and evaluation. In: Proc. of the 21st International Conference on Computational Linguistics and 46th Annual Meeting of the Association for Computational Linguistics (ACL/HLT), pp. 638–646. Columbus, Ohio, USA, June 2008
37. Rieser, V., Lemon, O.: Learning and evaluation of dialogue strategies for new applications: Empirical methods for optimization from small data sets. Computational Linguistics **37**(1), 2011
38. Rieser, V., Lemon, O.: Natural Language Generation as Planning Under Uncertainty for Spoken Dialogue Systems. In: Proc. of the Conference of European Chapter of the Association for Computational Linguistics (EACL), 2009
39. Rieser, V., Lemon, O.: Reinforcement Learning for Adaptive Dialogue Systems: A Data-driven Methodology for Dialogue Management and Natural Language Generation. Theory and Applications of Natural Language Processing, Springer (2011)
40. Schatzmann, J., Thomson, B., Weilhammer, K., Ye, H., Young, S.: Agenda-based user simulation for bootstrapping a POMDP dialogue system. In: Proc. of the North American Meeting of the Association of Computational Linguistics (NAACL), pp. 149–152. Rochester, New York, USA, April 2007
41. Schatzmann, J., Weilhammer, K., Stuttle, M., Young, S.: A survey of statistical user simulation techniques for reinforcement-learning of dialogue management strategies. Knowledge Engineering Review **21**(2), 97–126 (2006)
42. Stent, A., Walker, M., Whittaker, S., Maloor, P.: User-tailored generation for spoken dialogue: an experiment. In: Proc. of ICSLP, 2002
43. Stuttle, M.N., Williams, J.D., Young, S.: A framework for dialogue data collection with a simulated ASR channel. In: Proc. of the International Conference of Spoken Language Processing (Interspeech/ICSLP), Jeju, South Korea, October 2004
44. Sutton, R., Barto, A.: Reinforcement Learning. MIT Press (1998)
45. Thomson, B., Young, S.: Bayesian update of dialogue state: A POMDP framework for spoken dialogue systems. Computer Speech and Language **24**(4), 562–588 (2010)

46. Tvarožek, J., Bieliková, M.: Wizard-of-oz-driven bootstrapping of a socially intelligent tutoring strategy. In: Bastiaens, T., Ebner, M. (eds.) Proceedings of World Conference on Educational Multimedia, Hypermedia and Telecommunications 2011, pp. 3635–3644. Lisbon, Portugal, June 2011, AACE
47. Walker, M., Kamm, C., Litman, D.: Towards developing general models of usability with PARADISE. Natural Language Engineering **6**(3), 363–377 (2000)
48. Whittaker, S., Walker, M., Maloor, P.: Should i tell all? an experiment on conciseness in spoken dialogue. In: Proc. European Conference on Speech Processing (EUROSPEECH), 2003
49. Williams, J., Young, S.: Using Wizard-of-Oz simulations to bootstrap Reinforcement-Learning-based dialog management systems. In: Proc. of the 4th SIGDIAL Workshop on Discourse and Dialogue, pp. 135–139. Sapporo, Japan, July 2004
50. Young, S., Gasic, M., Keizer, S., Mairesse, F., Schatzmann, J., Thomson, B., Yu, K.: The Hidden Information State Model: a practical framework for POMDP-based spoken dialogue management. Computer Speech and Language **24**(2), 150–174 (2009)
51. Young, S.: Probabilistic methods in spoken dialogue systems. Philosophical Trans Royal Society (Series A) **358**(1769), 1389–1402 (2000)

Chapter 3
Data-Driven Methods for Spoken Language Understanding

James Henderson and Filip Jurčíček

Spoken dialogue systems need to be able to interpret the spoken input from the user. This is done by mapping the user's spoken utterance to a representation of the meaning of that utterance, and then passing this representation to the dialogue manager. This process begins with the application of automatic speech recognition (ASR) technology, which maps the speech to hypotheses about the sequence of words in the utterance. It is then the job of spoken language understanding (SLU) to map the word recognition hypotheses to hypothesised meanings. The representation of this meaning is called the semantics of the utterance.

To simplify the task of the SLU module in a SDS, the semantic representation is limited to those aspects of meaning that are required to perform the system's task effectively. Dialogue managers typically require semantics in the form of frames and slots, or dialogue acts, as described in [29, 31]. When more complex semantic representations are required, they can be defined by a grammar, such as the LR grammar for the GeoQuery domain [17].

Once we have defined the target semantics, we can develop an SLU module independently of the specific SDS by collecting and annotating a corpus of user utterances. As well as the GeoQuery [17] corpus, mentioned above, there are the ATIS [5], TownInfo [20], MEDIA [1], and LUNA [9] corpora, which also use a frame and slot semantics. Because the types of utterances, terminology, and target semantics vary from domain to domain, it is usually necessary to collect and annotate additional data for each new domain. Reducing the time and expense of this corpora creation is a common concern for SLU research. Several data-driven

J. Henderson (✉)
Département d'Informatique, Université de Genéve, Battelle, 1227 Carouge, Switzerland
e-mail: James.Henderson@unige.ch

F. Jurčíček
Faculty of Mathematics and Physics, Charles University in Prague, Malostranske namesti 25, 11800 Prague, Czech Republic
e-mail: jurcicek@ufal.mff.cuni.cz

O. Lemon and O. Pietquin (eds.), *Data-Driven Methods for Adaptive Spoken Dialogue Systems*, DOI 10.1007/978-1-4614-4803-7_3,

approaches to SLU try to reduce the amount of domain-specific data that is required to train an SLU model.

Another common concern for SLU research is the need to be robust to ill-formed utterances. SDS user utterances are spoken and spontaneous, so they differ systematically from the edited text typically used in NLP research. In particular, they include speech repairs, incomplete sentences and mistakes. In addition, the ASR may introduce mistakes. SLU systems need to extract the meaning that is available even from utterances that include such phenomena.

Once we have trained an SLU model for the domain, integrating it into the SDS imposes additional constraints on the design of the SLU module. Typically, the ASR module waits for the user to complete their utterance before sending the word recognition hypotheses to the SLU module. Meanwhile, the dialogue manager cannot compute how to respond to the user until it gets the semantics from SLU. In addition, there may be several recognition hypotheses for SLU to process, and SLU may want to hypothesise more than one semantics for each recognition hypothesis. For example, a typical N-best list of recognition hypotheses has N∼10 to 100 hypotheses. Therefore, an SLU module needs to be fast. Increasing the speed of computing the semantics from word recognition hypotheses is another common concern for SLU research.

Depending on the SDS architecture, the output of the SLU module may need to reflect the confidence that the hypothesised semantics accurately reflects what the user really said. In some cases, this may simply be the confidence score output by the ASR module for the associated word recognition hypothesis. In some cases, it may be a specific probability estimate, computed from the ASR confidence, the confidence of the SLU module, and whether multiple ASR hypotheses are mapped to the same semantics. For example, for the CLASSiC architecture, each hypothesised semantics is paired with an estimate of the likelihood of the speech stream given the semantics. This probability is estimated by marginalising out both the uncertainty from ASR and the uncertainty from SLU, for example, by running the SLU module on each ASR hypothesis and summing across equivalent semantics.

Most of the uncertainty about the semantics comes from uncertainty in the ASR hypotheses, but converting ASR confidence scores into probabilistic measures of this uncertainty can be difficult [28].

This chapter focuses on recent advances in data-driven SLU systems, with an emphasis on the models developed within the CLASSiC project. We present detailed explanations of three SLU systems developed in CLASSiC. First, Sect. 3.3 explains how transformation-based learning can be used to construct a very fast and robust parser. Second, Sect. 3.4 describes an efficient parsing method with state-of-the art results based on support vector machines (SVM). Third, Sect. 3.5 discusses how features derived from a domain-general semantic role labeller can be used in domains such as TownInfo. These systems provide good examples of approaches to the concerns mentioned above, including state-of-the-art performance, robustness to ill formed utterances, and handling small training sets. Before describing these systems, Sect. 3.1 outlines prior work on data-driven SLU, and Sect. 3.2 describes the semantic representation used in the TownInfo corpus, developed for the CLASSiC project. To conclude, Sect. 3.6 summarises the advantages and comments on future directions in data-driven SLU.

3.1 Prior Work

In addition to the systems described in detail in later sections, there has been a substantial amount of work on data-driven approaches to SLU. In this section, we outline some of the main contributions to this literature.

The hidden vector state (HVS) technique has been used to model an approximation of a pushdown automaton [12]. A vector state in the HVS parser represents a stack of a pushdown automaton, which keeps semantic information that spans over several words. Probabilistic stack shift operations represent state transitions. Given the limited number of possible operations, a robust transition model can be estimated using maximum-likelihood estimation (MLE). Finally, a deterministic algorithm is used to recover slot names and their values from the output sequence of vector states. The HVS model was evaluated on data from the ATIS domain which was converted into a frame based semantic annotation. The test extraction process is detailed in [13]. The authors reported a performance of 90.38% F-measure on identifying slot/value pairs. More recently, [35] presented discriminative training for the HVS model. In the ATIS domain, discriminative training of the HVS model gives a relative error reduction rate of 9 % when compared with MLE.

A probabilistic parser using combinatory categorical grammar (PCCG) has been used to map utterances to lambda-calculus expressions that represent logical form semantics [34]. Since typical combinatory categorical grammars are too rigid when processing spontaneous natural language, the work introduces new combinators in the grammar to handle spontaneous language. These new combinators, for example, allow for flexible word order or insertion of lexical items. In this work, the combinatory categorical grammar was generalised into a probabilistic model by learning a log-linear model. The features used in the model include the number of times each combinator was used, the number of used lexical items from the parser's lexicon, and features representing the output logical form. The online learning algorithm uses perceptron updates to adjust weights of features in the log-linear model. The evaluation was performed on the ATIS and Geo880 corpora. On the ATIS corpus, this technique produces state-of-the-art performance of 95.9 % F-measure. However, in addition to using the lexical categories (city names, airport names, etc) readily available from the ATIS corpus, this method also needs a considerable number of handcrafted entries in its initial lexicon.

Markov logic networks (MLN) have been used to extract slot values by combining statistical and logical reasoning [23]. In this approach, first-order clauses are used to represent two types of relationships: (1) relations between slot names and their values and (2) relations between slots themselves. By attaching weights to each of the clauses, a probabilistic extension is obtained. Such weighted clauses are used as templates for features of Markov networks, where standard methods for probabilistic inference can be used. In this work, the MLN model was evaluated on the ATIS corpus and the reported performance was 92.99 % F-measure.

Dinarelli et al. [8] train SLU models on the ATIS, MEDIA, and LUNA corpora. They show that local n-gram based SLU models can be improved using discriminative reranking with global features[1]. For the local models, they experiment with finite-state transducers, SVM and CRFs. The global features are manifested in kernels designed specifically for the semantic structures output by such SLU models. With ASR input, they achieve 13.2 % CER on Atis, 26.3 % CER on MEDIA, and 31.4 % CER on LUNA.

Wu et al. [32] develop an SLU system that first classifies an utterance according to its topic [such as its dialogue act type (DAT)], and then finds a set of slot-value pairs appropriate for that topic. The latter classifier uses local features, but will reclassify values if there is a conflict between slot–value pairs. To reduce the need to annotate training data, they use a combination of active learning and self-training for the topic classifier, and exploit constraints between slot–value pairs in a self-training algorithm for the slot–value classifiers.

A comprehensive study of different models for SLU is presented in [10]. In this work, six parsing methods including finite state transducers (FSTs), statistical machine translation (SMT), maximum entropy Markov models (MEMMs), support vector machines (SVMs), conditional random fields (CRFs) or dynamic bayesian networks (DBNs) were compared. The evaluation was performed on three corpora including the French MEDIA corpus and the Polish and Italian LUNA corpora. Each model was evaluated on manual transcriptions as well as ASR hypotheses. In addition to individual systems evaluation, the work tested the ROVER method for system combination. Among all the evaluated methods, the model based on CRFs was the best performing. On the MEDIA corpus, the CRF model achieved 12.6 % CER. When the ROVER method with all six systems was applied, 12.0 % CER was obtained.

3.2 Semantic Representation in the TownInfo Corpus

The semantic representation used in the TownInfo corpus is typical of the kinds of frame and slot semantic representations used in SDS. The TownInfo corpus was developed in the CLASSiC project. It consists of tourist information dialogues in a fictitious town. The dialogues were collected through user trials in which users searched for information about a specific venue by interacting with a dialogue system in a noisy background.

---

[1] Similar experiments are reported in [6], where generative models are reranked with discriminative models, and in [7], where they focus on the advantages of discriminative reranking for small datasets.

Table 3.1 An example of a dialogue with the semantic annotation of the input utterances from the TownInfo corpus

System	Thank you for calling the TownInfo system
User	Hi, I'm looking for a cheap Chinese restaurant
	`inform(type=restaurant,food=Chinese,pricerange=cheap)`
System	Yu Garden serves Chinese food. It is in the cheap price range
User	Ok, give me the address
	`request(addr)`
System	Yu Garden is located at 529 Newmarket Road
User	Thank you. Goodbye
	`bye()`

In the TownInfo corpus, the input utterances are mapped into a semantic representation in the form of dialogue acts. A dialogue act is composed of a DAT and a list of slot–value pairs (e.g. type=hotel, near=market square). The DAT typically conveys the user intention, such as inform, request, etc. The slot-value pairs specify arguments to the DAT. In some cases, the value can be omitted, for example, where the intention is to query the value of a slot, as in "request(food)". For example, the utterance "I want an Indian restaurant in the cheap price range" is semantically represented as

```
inform(food=Indian, type=restaurant, pricerange=cheap)
```

A full description of the dialogue act set used by the CLASSiC project is given in [33]. An example of a dialogue using this semantic representation is shown in Table 3.1.

3.3 Using Transformation-Based Learning for SLU

This section describes a semantic parser based on transformation-based learning (TBL) [2]. The TBL parser transforms an initial semantic hypothesis into the correct semantics by applying an ordered list of transformation rules [16]. In each iteration, a transformation rule corrects some of the remaining errors in the semantics. To handle long-range dependencies between words, the parser uses features extracted from dependency parse trees provided by the RASP syntactic parser [3]. The transformation rules are learnt automatically from a training corpus with no prior linguistic knowledge and no alignment between words and semantic concepts. The learning algorithm produces a compact set of rules which enables the parser to be very efficient while retaining high accuracy.

Each transformation rule is composed of a trigger and a transformation. The trigger is matched against both the utterance and the semantic hypothesis, and when successfully matched, the transformation is applied to the current hypothesis. In the TBL parser, a trigger contains one or more conditions as follows: the utterance contains N-gram, the DAT equals D, and the semantics contains slot S. If a trigger contains more than one condition, then all conditions must be satisfied. N-gram triggers can be unigrams, bigrams, trigrams, or skipping bigrams which can skip up to three words. A transformation performs one of the following operations: replace the DAT, add a slot, delete a slot, and replace a slot. A replacement transformation can replace a whole slot, a slot name, or a slot value. Some example rules with triggers composed of unigrams, skipping bigrams, and DAT matching are:

Trigger	Transformation
"I want"	Replace DAT by "inform"
"can * give" & DAT=inform	Replace DAT by "request"
"cheap"	Add the slot "pricerange = cheap"
"centre"	Add the slot "area = centre"
"near"	Replace the slot "area = *" by "near = *"

The first rule replaces the DAT by "inform" if the bigram "I want" is in the utterance. The second rule changes DAT from "inform" to "request" if the utterance contains the words "can" and "give", which can be up to three words apart. The fourth rule adds the slot "area=centre" whenever the utterance contains the word "centre". Finally, every slot name "area" is replaced by "near" if the utterance includes the word "near".

3.3.1 Example of Parsing

This section demonstrates the parsing process on the example *"i am at the west side shopping centre could you tell me a nearby hotel"* from the TownInfo domain.

First, the DAT "inform" and no slots are used as the initial semantics. As a result, the initial semantics is as follows:

DAT = inform

Second, the rules, whose triggers match the utterance and the hypothesised semantics, are sequentially applied.

#	Trigger	Transformation
1	"Hotel"	Add the slot "type=hotel"
2	"Centre"	Add the slot "area=centre"

Use the previous transformations results in the following semantic hypothesis:

DAT = inform
type = hotel
area = centre

As the word "centre" is associated with the "area" slot most of the time in the TownInfo dataset, the parser learns to associate it with the "area" slots. The incorrect classification of the word "centre" is a result of such a generalisation. However, the TBL method learns to correct its errors. Therefore, the parser also applies the error correcting rules at a later stage. For example, the following rule corrects the incorrect inclusion of the slot "area=centre".

#	Trigger	Transformation
3	"West side shopping" & area=centre	Replace the slot "area=centre" by "near=west side shopping"

In this case, the incorrect slot is replaced by correct interpretation to produce the following semantic hypothesis:

DAT	=	inform
type	=	hotel
near	=	west side shopping

3.3.2 Improving the Disambiguation of Long-Range Dependencies

Besides simple n-grams and skipping bigrams, more complex lexical or syntactic features can be used. Kate [18] used manually annotated dependency trees to capture long-range relationships between words. In a dependency tree, each word is viewed as the dependant of one other word, with the exception of the root. Dependency links represent grammatical relationships between words. Kate showed that word dependencies significantly improve semantic parsing because long-range dependencies from an utterance tend to be local in a dependency tree. Instead of using manually annotated word dependencies, the TBL method uses automatically generated dependencies provided by the RASP dependency parser [3].

3.3.3 Learning

The main idea behind TBL [2] is to learn an ordered list of rules which incrementally improve an initial semantic hypotheses. Note that the list of rules must be ordered because each learnt rule corrects some of the remaining errors after applying the preceding rules. The initial assignment is made based on simple statistics–the most common dialogue act is used as initial semantics. The learning is conducted in a greedy fashion, and at each step, the algorithm chooses the transformation rule that reduces the largest number of errors in hypotheses. Errors include DAT substitutions, slot insertions, slot deletions, and slot substitutions. The learning process stops when the algorithm cannot find a rule that improves the hypotheses beyond some preset threshold.

As in previous work [13, 23, 34], the parsers relies on a database characterising entities of interest in the dialogue domain as category/value pairs (e.g.

`PLACE_NAME = main square`). After parsing, a deterministic algorithm recovers the original values for category labels, which is detailed in [20].

The learning algorithm consists of the following steps:

- **Input**: a set of (utterance, semantic tree) pairs, a domain database
- **Output**: a classifier for the semantic tree of the input utterance

1. Substitute database values in the training utterances with category labels.
2. Assign initial semantics to each utterance.
3. Repeat as long as the number of errors on the training set decreases:
   (a) Generate all rules which correct at least one error in the training set.
   (b) Measure the number of errors corrected minus the number of errors introduced by each rule.
   (c) Select the rule with the largest number of corrected errors.
   (d) Stop if the number of corrected errors is smaller than threshold T.
   (e) Add the selected rule to the end of the rule list and apply it to the current state of the training set.

To limit overfitting the training data, rules inferred at the end of the learning are pruned. To determine which rules should be pruned, the learning algorithm sequentially applies each rule on the development set and measure the number of errors it produces. The algorithm then retains only the first N rules for which the TBL parser gets the lowest number of errors on the development set.

3.3.4 Parsing Efficiency

The learning time of the TBL parser is acceptable, and the parsing process is very efficient. First, the learning time is about 24 h on an Quad Core Intel Pentium 2.8 GHz for each dataset. In the TownInfo domain, the TBL parser generates about 1 M potential transformation rules in each iteration; however, only a fraction of these rules have to be tested because the search space can be efficiently organised [2]. Second, the TBL parser is able to parse an utterance in 6 ms, on average. The TBL parser is very efficient on domains such as TownInfo because the final list of learnt rules is small. In the TownInfo dataset, there are 14 DATs and 14 slots, and the total number of learnt rules is 195. This gives 6.9 rules per semantic concept, on average.

3.4 The Semantic Tuple Classifiers Model of SLU

The semantic tuple classifiers (STC) method is an efficient technique that learns discriminative semantic concept classifiers whose outputs are used to construct a semantic interpretation of the input utterance in the form of a semantic tree [20]. In this approach, concept classifiers predict fragments of the semantic representation using SVM and n-gram lexical features. By recursively calling the SVM classifiers,

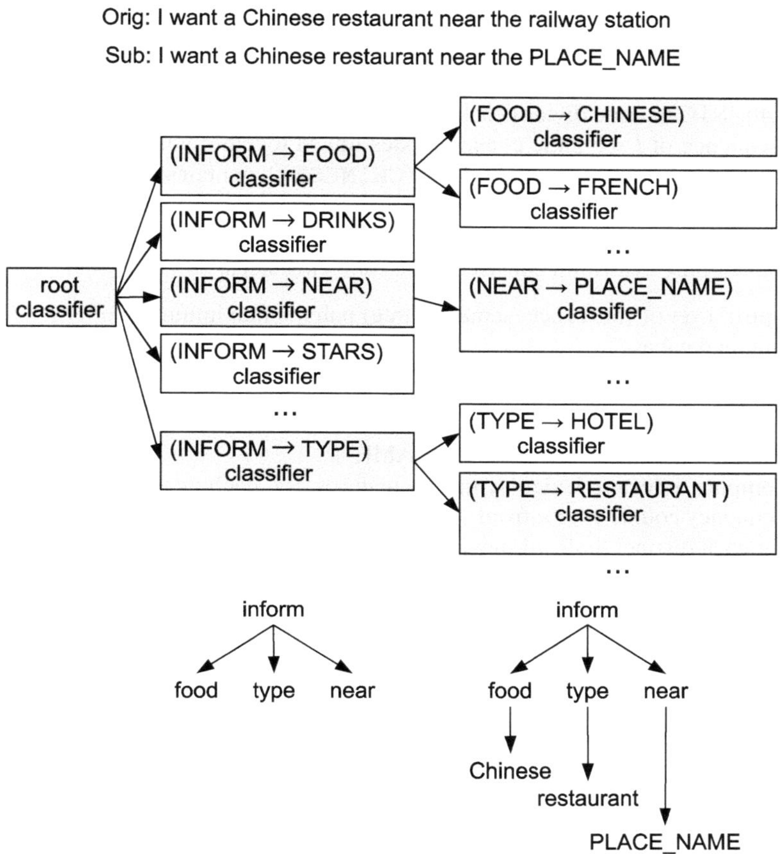

Fig. 3.1 Semantic tree derivation for an utterance from the TownInfo dataset with a tuple length of 2, with positive tuple classification in darker boxes

a complete semantic tree can be built. Although the STC method can produce arbitrary large parse trees, this is not often necessary. For the TownInfo corpus, and the CLASSiC project, STC only needs to produce semantic trees with depth limited to three. The root of each semantic tree represents the DAT, its children represent the slot names, and the leaves represent slot values.

3.4.1 Training the STC Classifiers

The input of the learning algorithm presented in this section is a set of utterances and their corresponding semantic trees. As in the TBL model, the parser uses a database with category labels and their possible values. While this approach helps to scale to a large set of possible concept values, it can lead to lower parsing performance for concepts with just a few values (e.g. the `TYPE` attribute in Fig. 3.1

is only associated with RESTAURANT and HOTEL). Therefore, the STC model specifies which category values are replaced by their category names. To control the accuracy of semantic parsing and to generalise over possible unseen semantic trees, the STC method divides each semantic tree into concept tuples, which consist of a sequence of l semantic concept nodes linked together within a single branch. For example, the tree `INFORM(FOOD(CHINESE))` contains two tuples of length 2 (i.e., `INFORM`→`FOOD` and `FOOD`→`CHINESE`) and one tuple of length 3 (i.e. `INFORM`→`FOOD`→`CHINESE`).

The learning algorithm consists of the following steps:

- **Input**: a set of (utterance, semantic tree) pairs, a maximum tuple length l and a domain database
- **Output**: a set of STC and a domain grammar

1. Substitute database values in the training utterances with category labels (e.g., "I want a restaurant near PLACE_NAME").
2. Compute relevant lexico-syntactic features for each utterance (e.g. n-gram frequency counts with n from 1 to 3).
3. For each distinct tuple, of maximum l concepts, in the semantic trees:
   (a) Create a dataset associating each training utterance with a binary class representing whether the tuple occurs in the utterance's semantic tree.
   (b) Train a binary semantic tuple classifier (STC) that predicts the class—i.e. the tuple—from the feature values. The root concept (e.g. DAT) is predicted using a single multi-class classifier.
4. Construct a domain grammar that matches all trees in the training set.

Concerning the learning algorithm of individual STC, [20] used SVM classifiers with linear kernels. Although any set of lexical, syntactic, and semantic features can be used to train the classifiers, the STC model only used n-gram frequency counts computed over the training utterances. The maximum n-gram size ($1 \leq n \leq 3$) as well as the SVM misclassification cost parameters are optimised individually for every tuple classifier, by performing cross-validation on the training data.

3.4.2 Decoding User Utterances

Once the STC classifiers have been trained, they can be used to predict the meaning of unseen user utterances. Figure 3.1 illustrates the parsing process for the TownInfo domain, in which the semantic representation is reconstructed by combining positive semantic tuple classes.

The STC method has two variants: (1) high-precision mode and (2) high-recall mode. In the high-precision mode, the algorithm relies only on the outputs of the classifiers to expand the tree (e.g., `FOOD`→`CHINESE` can only be appended if the `FOOD` concept is already part of the tree), whereas in high-recall mode, the tree can be expanded as long as the result matches the domain grammar (e.g., `FOOD→CHINESE` can be appended to `INFORM` if this combination has been seen during training).

Since [20] reported that the parser performs significantly better in the high recall mode, only this variant is described. The decoding algorithm for high recall mode consists of the following steps:

- **Input**: the user utterance, the semantic concept classifiers, the domain grammar and database
- **Output**: the semantic tree of the input utterance

1. Substitute database values in the utterance with category labels.
2. Compute the utterance's n-gram features and filter out those not seen during training (e.g. unseen n-grams).
3. Run all STC on the utterance's features and set T equal to the set of positively classified tuples.
4. Initialise the output semantic tree as the predicted root, and a variable r pointing to the root concept.
5. For each tuple t of T whose root has the same concept as r, append t's non-root nodes to r in the semantic tree, and remove t from T. Start over recursively by setting r equal to t's terminal node.
6. For each remaining tuple t of T that is dominated by the concept r in the domain grammar, append t to r in the semantic tree. Start over recursively by setting r equal to t's terminal node.
7. Associate each of the tree's terminal nodes corresponding to a database category label with the corresponding value in the utterance (e.g. `PLACE_NAME` becomes railway station).

The output of this algorithm is a semantic tree of the utterance, with terminal concepts associated with database values.

3.4.3 Parsing Efficiency

The method scales to large semantic representations. Since the STC method requires C^l classifiers, where C is the number of possible semantic concepts and l the tuple length, on the input utterance, the computational cost of our approach at run-time is limited by the number of distinct tuples in the training data. Using approximately 250 linear support vector machine classifiers on a Quad Core Pentium 2.4 GHz, an utterance can be parsed in less than 200 ms on average in the TownInfo domain.

Although the STC method is slower than the TBL algorithm, the STC method is still suitable for real-time dialogue systems. Moreover, it is possible to use the STC method to calculate probabilities for multiple semantic hypothesis given the input utterance, as described in [19].

Table 3.2 Results on the ATIS test set

	Act type	Slot-Value		
	Acc	Prec	Rec	F
	Transcribed utterances			
Parser PCCG	–	95.11	96.71	95.90
Parser HVS	–	–	–	90.30
Parser MLN	–	–	–	92.99
Parser TBL	92.82	96.37	95.12	95.74
Parser STC	92.63	96.73	92.37	94.50

Table 3.3 Results on the TownInfo test sets for both transcribed utterances and noisy ASR output

	Act Type	Slotvalue		
	Acc	Prec	Rec	F
	Transcribed utterances			
Parser Phoenix	94.82	96.33	94.22	95.26
Parser TBL	94.90	96.05	94.66	95.35
Parser STC	94.92	97.39	94.05	95.69
	ASR output			
Parser Phoenix	74.63	90.28	79.49	84.54
Parser TBL	83.24	92.72	83.42	87.82
Parser STC	85.15	94.03	83.73	88.58

3.4.4 Evaluation

In this section, the STC and TBL parsers are evaluated on two distinct corpora, and compared with state-of-the-art techniques and a handcrafted Phoenix parser [30].

In order to compare the results with previous work [13, 23, 34], the methods are applied to the ATIS dataset [5]. The ATIS corpus is divided into training (5,012 utterances), development (445 utterances), test (448 utterances) datasets. The evaluation criterion is the F-measure of the number of reference slot/value pairs that appear in the output semantics (e.g. from.city = New York). He and Young [13] detail the test data extraction process.

The methods are also evaluated on the TownInfo dataset. The TownInfo training, development, and test sets, respectively, contain 8,396, 986, and 1,023 transcribed utterances. The data includes the transcription of the top hypothesis of a speech recogniser, which allows us to evaluate the robustness of our models to recognition errors (word error rate = 34.4 %). The TBL and STC parsers are compared to the handcrafted Phoenix parser [30]. The Phoenix parser implements a partial matching algorithm that was designed for robust SLU.

The results for the two datasets are shown in Tables 3.2 and 3.3. The model accuracy is measured in terms of precision, recall, and F-measure (harmonic mean of precision and recall) of the slot/value pairs. Both slot and value must be correct to count as a correct classification.

Results on the ATIS dataset show that both the TBL (F-measure = 95.74 %) and STC (F-measure = 94.54 %) parsers are competitive with respect to the Zettlemoyer & Collins' PCCG model [34] (95.90 %). Note that this PCCG model makes use of a considerably large number of handcrafted entries in their initial lexicon. In addition, the TBL and STC models outperform the HVS [13] and MLN [23] parsers. Concerning the TownInfo dataset, Table 3.3 shows that both TBL and STC outperform the baseline Phoenix parser on the ASR output. Regarding the performance on the transcribed output, the performance of all three parsers is similar and no significant differences can be observed.

3.5 Using Domain-General Semantic Role Labelling in SLU

One way to reduce the amount of domain-specific data that is required to train an SLU model is to take advantage of domain-general data on semantic interpretation. For SLU systems, such as those discussed above, where natural language utterances are mapped directly into the required frame and slot semantics, the data they are trained on needs to cover both the range of variability in the semantics and the range of variability in the natural language expressions that convey that semantics. Because the set of frames and slots is domain specific, this data must be domain specific. But in general, the variability in natural language expressions is not domain specific, so we should be able to reduce the amount of domain-specific data that is required if we first normalise the natural language to a domain-general semantic representation, and then map to the domain-specific semantics.

In this section, we present work on using semantic role labelling (SRL) as a form of domain-general semantic preprocessing to simplify the mapping to domain-specific semantics in SLU. SRL explicitly identifies some kinds of phrases, such as locatives and verbal arguments, which are important for determining the basic meaning of an utterance. However, the existing corpora annotated with SRLs are in the domain of edited text (e.g. see the CoNLL 2008 and 2009 shared tasks [11, 27] or the FrameNet corpus [26]). Edited text differs systematically from the spoken utterances of SLU for SDS.

3.5.1 SRL on SDS Utterances

To show that SRL can be useful for SDS, it is first necessary to show that SRL parsers can be applied effectively to spoken utterances. Much of the work on SRL for SLU has actually only addressed this concern [4, 25]. Coppola et al. [4] train a FrameNet semantic role classifier on a small corpus of human–human task-oriented dialogues and achieves 76 % F_1 on role identification and labelling. This is a usable level of accuracy despite the small corpus size of only 1,677 annotated frame instances, but this level was achieved using gold annotations for predicate identification, frame labelling, and syntactic structure.

Van der Plas et al. [25] trained a complete SRL system to produce PropBank annotations for the kinds of spoken utterances found in the TownInfo corpus.

Utterance: *in the east*

```
Parse:    (TOP (FRAG
                (PP@AM_LOC
                  (IN@AM_LOC in)
                    (NP (DT the)
                      (NNP east) ) )
                (. .) ) )
```

Features: localslot=area, localvalue=area-east,
word=in, tagword=in_IN@AM_LOC, bigram=^_in,
label=IN@AM_LOC, in_IN@AM_LOC_parent=PP@AM_LOC,
in_IN@AM_LOC_rgtsib=NP, in_IN@AM_LOC_maxp_parent=FRAG,
in_IN@AM_LOC_maxp_rgtsib=., in_IN@AM_LOC_maxp_rgtsibwrd=.,
word=east, tagword=east_NNP bigram=the_east
label=NNP, east_NNP_parent=NP, east_NNP_lftsib=DT,
east_NNP_lftsibwrd=the, east_NNP_maxp_parent=PP@AM_LOC,
east_NNP_maxp_lftsib=IN@AM_LOC, east_NNP_maxp_lftsibwrd=in

Output: `[inform(area=east), 0.973584]`

Fig. 3.2 An example of a structure output by the syntactic-semantic parser, the features computed from it, and the dialogue act and probability that these features map to. Semantic role labels are those that follow "@"

Rather than annotating spoken dialogue data, they generated an appropriate artificial corpus from the existing PropBank corpus of edited text, and trained an SRL parser on the artificial corpus.

Van der Plas et al. [25] used the syntactic-semantic parsing method proposed by [22], which produces a merged representation of syntactic constituency trees from the Penn Treebank [21] annotated with SRL information from PropBank [24]. An example of this output is given in Fig. 3.2. Note that the phrase is labelled as a locative modifier, even though the verb that is modified is not present in this fragment. This is a statistical parsing model, which uses the neural network parsing architecture of [14] to estimate its probabilities. The output of the parser is a list of the most probable parses, with their generative probabilities. A form of beam search is used to search for the most probable parses, which allows a trade-off between speed and accuracy by varying the beam width.

Van der Plas et al. [25] first applied the parser trained only on PropBank to the utterances from the TownInfo domain. They found that this parser handled many differences with written text robustly. However, it had trouble with sentence fragments (e.g. *"in the east"*), which are very rare in PropBank but very common in system-driven dialogues like those in TownInfo. They addressed this problem by retraining the parser on a new corpus where they added artificially generated sentence fragments.

To ensure that the distribution in the artificially generated corpus is as close as possible to that found in the TownInfo corpus, [25] used a three-component model to generate the artificial corpus. The first component is a theory of the types of constructions found in the domain, which were characterised primarily as fragments rooted in a specific constituent category (e.g. NP, noun phrase) and all words within that constituent. The second component is the distributions of these types (e.g. NPs versus full sentences) that are found in the TownInfo corpus. This necessitated the annotation of a subset of TownInfo utterances with these type labels but not with full parses. The third component is the distribution of parse structures found in the original semantically annotated corpus. The first component defined rules for extracting fragments from this corpus, keeping their internal parse structures, thereby keeping the distribution from the third component for these internal structures. The distribution from the second component was then used to sample from this collection of fragments (and full sentences) to produce the artificial corpus, thereby generating a corpus with the distribution of fragments found in the TownInfo data.

An SRL parser trained on this artificial dataset achieved 79.5 % F_1 on heldout TownInfo data, only 1.2 % worse than the parser trained and tested on the PropBank corpus, and 6.6 % better than the parser trained on the PropBank corpus and tested on the TownInfo corpus [25].

3.5.2 SLU with SRL Features

The above SRL parser of [25] was later incorporated into a complete SLU system [15]. This system includes a preprocessor, the SRL parser, and a model for identifying the output dialogue act. The decoding algorithm for this system consists of the following steps:

- **Input**: the user utterance, the SRL parser, and the dialogue act classifier
- **Output**: the dialogue act of the input utterance

1. Preprocess the user utterance to make it easier to parse.
2. Run the SRL parser on the preprocessed utterance, producing the set of parses whose probability is sufficiently close to that of the best parse.
3. For each parse, extract a set of features from the parse and the utterance.
4. For each parse, run the dialogue act classifier on the extracted features, producing (a distribution over) the most probable dialogue act(s).
5. Sum the probabilities of the parse-act pairs with the same dialogue act, producing (a distribution over) the most probable dialogue act(s).

The robustness of the parser meant that the preprocessing could be reduced to just removing filler words, such as "um". The SRL parser was the above syntactic-semantic parser trained on an artificial dataset including sentence fragments [25]. Two models were developed for identifying the output dialogue act, both based on the semantic tuple classifier [20] discussed in Sect. 3.4.

One advantage of the STC model is that it has a simple interface with the input utterance, namely, the vector of features that is input to all the statistical classifiers. This allows the output of the SRL parser to be incorporated into an STC model by simply converting the syntactic-semantic structure into a vector of

Table 3.4 The features computed from a parse

Feature name	= Feature value
`localvalue`	String found which matches a database entry
`localslot`	Slot for `localvalue`
`word`	Word string
`label`	POS-tag
`tagword`	Pairing of word with POS-tag
`norm`	Normalised word, if different from `word`
`bigram`	Pair of adjacent words
`w_t_parent`	Label of parent of `w_t`, a word-tag pair
`w_t_lftsib`	POS-tag/label of left sibling of `w_t`
`w_t_rgtsib`	POS-tag/label of right sibling of `w_t`
`w_t_lftsibwrd`	Word (if any) that is left sibling of `w_t`
`w_t_rgtsibwrd`	Word (if any) that is right sibling of `w_t`
`w_t_maxp_parent`	Label of parent of `w_t_maxp`, the maximal projection of parent of `w_t`
`w_t_maxp_lftsib`	POS-tag/label of left sibling of `w_t_maxp`
`w_t_maxp_rgtsib`	POS-tag/label of right sibling of `w_t_maxp`
`w_t_maxp_lftsibwrd`	Word (if any) that is left sibling of `w_t_maxp`
`w_t_maxp_rgtsibwrd`	word (if any) that is right sibling of `w_t_maxp`
`w_t_vgov`	POS-tag of governing verb of `w_t`
`w_t_vgovwrd`	Word of governing verb of `w_t`
`w_t_vgovroot`	Label of maximal projection of governing verb of `w_t`

Table 3.5 Results on the TownInfo test sets for both transcribed utterances and noisy ASR output

	Act type	Slotvalue		
	Acc	Prec	Rec	F
	Transcribed utterances			
Parser MaxEnt	93.55	94.37	85.66	89.80
Parser SVM	94.23	97.20	92.75	94.93
N-gram SVM (STC)	94.92	97.39	94.05	95.69
	ASR output			
Parser MaxEnt	82.69	89.71	73.94	81.07
Parser SVM	82.60	90.97	79.18	84.67
N-gram SVM (STC)	85.15	94.03	83.73	88.58

semantically informative features. A set of patterns were developed for extracting semantically informative features from the output of the SRL parser. These include lexical features, and some bi-lexical features. The set of features is summarised in Table 3.4. Examples of these features are given in Fig. 3.2.

These features were used as input to two different dialogue act classifiers. First, the dialogue act classifier developed for the STC model by [20] was applied to the SRL feature set. Results for this model are shown below as "Parser SVM" in Table 3.5. Maximum Entropy classifiers were also used. These classifiers directly provide a probability distribution over possible interpretations, but these classifiers do not work as well as SVMs with small datasets. Results for this model are shown below as "Parser MaxEnt" in Table 3.5.

The Maximum Entropy dialogue act classifier decomposes the dialogue act into one classifier to choose the DAT (e.g. `inform` versus `request`), and separate classifiers for choosing each node in the semantic tree. These classifiers compute the probability of each node conditioned on its parent node, so that the semantic tree is generated by a probabilistic context-free grammar. This ensures that the score for each dialogue act is a valid probability. The semantic tree is generated as follows. If the DAT takes no arguments, then stop. If the DAT takes a special first argument (e.g. `request`), then choose that argument. Then for each possible slot name, first choose whether it is absent, positive ("="), or negative ("!="). Then choose its value. There is the option of generating the value as a specific string (e.g. "restaurant"), or as a database type (e.g. "name") which is subsequently substituted with a string found in the utterance (e.g. "Alexander Hotel").

3.5.3 Evaluation of SRL for SLU

The accuracies of the models discussed above on the TownInfo test sets are shown in Table 3.5. The top half of the table shows accuracies for training and testing on transcribed utterances, and the bottom half shows accuracies for training and testing on the noisy output of an ASR module. For these tests, only the single best hypothesis from the ASR module was used, not the n-best hypotheses.

For the models which exploit an SRL parser, the parser was trained with a small vocabulary (813 tag-word pairs) and did decoding with a small beam width (pruning to five analyses after each word). These choices made the parser relatively fast, resulting in the full SLU module processing 30 words per second. But it also resulted in a less accurate parser.

Between the data-driven models which exploit an SRL parser, there is a clear improvement in using the SVM classifier ("Parser SVM") over using the MaxEnt classifier ("Parser MaxEnt"), in particular on identifying slot–value pairs. This is probably due to the small number of positive examples for many of the slot-value classifiers, as apposed to the relatively large number of data points for DATs. SVMs perform particularly well with small datasets. It could also be due to the different ways individual slot/value classifiers are combined in the two systems, particularly because the high-recall version of the SVM classifier was used.

Comparing the results using the SVM classifier with SRL-parser-derived features ("Parser SVM") to the same classifier with only n-gram features [20] ("N-gram SVM"), very similar results are achieved on transcribed utterances, but the n-gram features perform better on noisy ASR output. This is not an entirely fair comparison, because more fine-tuning of meta-parameters was done for the n-gram features, but the difference on ASR output data probably cannot be accounted for in this way. This difference suggests that the current parser does not perform as well on noisy input. This limitation might be somewhat mitigated if the evaluations were done using n-best ASR output, rather than only having access to the one-best ASR output as here. Nonetheless, in future work, it may be useful to train a syntactic-semantic parser on artificial treebank data where words have been substituted using a confuser that simulates ASR errors.

We consider that the equivalent performance of parser-derived features and n-gram features on transcribed utterances is a promising initial result. The types of phenomena where we would expect semantic parsing to be of help, such as the scope of negation (e.g. "not French or Chinese" versus "not French, Chinese"), are very rare in the TownInfo corpus. So the fact that we are doing well even with utterances that are simple enough for n-grams to give state-of-the-art results suggests that semantic parsing should result in an improvement as we move to more sophisticated dialogue systems with more natural interactions and better quality speech recognisers.

3.6 Conclusion

Data-driven approaches have produced several state-of-the-art systems for SLU. The robustness of these approaches allows them to handle the ill-formed language typical of spontaneous spoken utterances, as well as the errors introduced by current ASR technology. Since many data-driven approaches only require very shallow language processing and simple classifiers, they tend to also have the speed necessary to support the real-time interactions of spoken dialogue systems. One limitation of such shallow direct methods is that they rely on sufficient quantities of domain-specific SLU training data, which can be slow and costly to collect and annotate. Several data-driven SLU methods address this issue, including work on using domain-general SRL parsers to simplify the mapping to the domain-specific semantics.

When dealing with spoken utterances, it is inevitable that the interpretation of some utterances will be uncertain, particularly given the difficulty of ASR. An additional advantage of data-driven methods is that they allow us to quantify this uncertainty in terms of probabilities and pass this information on to the dialogue manager. This gives the dialogue manager information about both the range of interpretations and the level of uncertainty, putting it in the best possible position to make intelligent dialogue management decisions.

While robustness and speed still remain challenges, research on SLU is looking towards domains with richer semantic representations and more complex linguistic constructions. This trend makes the problem of small datasets even more acute, but it also increases the mutual relevance with text-based semantic processing research. The complexity of broad-coverage deep semantic interpretation and the specific challenges of SLU make this a very challenging goal, but it is necessary to achieve the natural conversational interactions with spoken dialogue systems that we will expect in the future.

Acknowledgements This work was partially funded by the EU FP7 Programme under grant agreements 216594 (CLASSiC), 287615 (PARLANCE) and 249119 (META-NET), by Ministry of Education, Youth and Sports of the Czech Republic under the grant agreement LK11221, and by core research funding of Charles University in Prague. The authors would like to thank François Mairesse, for discussions about the STC parser, and Lonneke van der Plas and Paola Merlo for discussions about the work on SRL for SLU.

References

1. Bonneau-Maynard, H., Ayache, C., Bechet, F., Denis, A., Kuhn, A., Lefvre, F., Mostefa, D., Qugnard, M., Rosset, S., Servan, J., Vilaneau, S.: Results of the French Evalda-Media evaluation campaign for literal understanding. In: Proceedins of the International Conference on Language Resources and Evaluation (LREC), pp. 2054–2059, 2006
2. Brill, E.: Transformation-based Error-driven Learning and natural language processing: A case study in Part-of-Speech Tagging. Computational Linguistics **21**(4), 543–565 (1995)
3. Briscoe, E., Carroll, J., Watson, R.: The second release of the RASP system. In: Proceedings of COLING/ACL, 2006
4. Coppola, B., Moschitti, A., Riccardi, G.: Shallow semantic parsing for spoken language understanding. In: Proceedings of Human Language Technologies: The 2009 Annual Conference of the North American Chapter of the Association for Computational Linguistics, Companion Volume: Short Papers, pp. 85–88, 2009
5. Dahl, D.A., Bates, M., Brown, M., Fisher, W., Hunicke-Smith, K., Pallett, D., Pao, C., Rudnicky, A., Shriberg, E.: Expanding the scope of the ATIS task: The ATIS-3 corpus. In: Proceedings of the ARPA HLT Workshop, 1994
6. Dinarelli, M., A. Moschitti, Riccardi, G.: Re-ranking models for spoken language understanding. In: Proceedings of the 12th Conference of the European Chapter of the ACL (EACL 2009), pp. 202–210, 2009
7. Dinarelli, M., Moschitti, A., Riccardi, G.: Re-ranking models based-on small training data for spoken language understanding. In: Proceedings of the 2009 Conference on Empirical Methods in Natural Language Processing, pp. 1076–1085, 2009
8. Dinarelli, M., Moschitti, A., Riccardi, G.: Discriminative reranking for spoken language understanding. IEEE Transactions on Audio, Speech, and Language Processing **20**(2), 526 –539 (2012)
9. Dinarelli, M., Quarteroni, S., Tonelli, S., Moschitti, A., Riccardi, G.: Annotating spoken dialogs: From speech segments to dialog acts and frame semantics. In: Proceedings of SRSL 2009, the 2nd Workshop on Semantic Representation of Spoken Language, pp. 34–41, 2009
10. Hahn, S., Dinarelli, M., Raymond, C., Lefevre, F., Lehnen, P., De Mori, R., Moschitti, A., Ney, H., Riccardi, G.: Comparing Stochastic Approaches to Spoken Language Understanding in Multiple Languages. IEEE Transactions on Audio, Speech, and Language Processing **19**(6), 1569–1583 (2011)
11. Hajič, J., Ciaramita, M., Johansson, R., Kawahara, D., Martí, M., Màrquez, L., Meyers, A., Nivre, J., Padó, S., Štěpánek, J., Straňák, P., Surdeanu, M., Xue, N., Zhang, Y.: The conll-2009 shared task: Syntactic and semantic dependencies in multiple languages. In: Proceedings of the Thirteenth Conference on Computational Natural Language Learning (CoNLL 2009): Shared Task, pp. 1–18. Boulder, Colorado, June 2009
12. He, Y., Young, S.: Hidden vector state model for hierarchical semantic parsing. In: Proceedings of ICASSP, Hong Kong (2003)
13. He, Y., Young, S.: Semantic processing using the Hidden Vector State model. Computer Speech & Language **19**(1), 85–106 (2005)
14. Henderson, J.: Inducing history representations for broad coverage statistical parsing. In: Proc. joint meeting of North American Chapter of the Association for Computational Linguistics and the Human Language Technology Conf., pp. 103–110. Edmonton, Canada (2003)

15. Henderson, J.: Semantic decoder which exploits syntactic-semantic parsing, for the towninfo task. Technical Report Deliverable 2.2, CLASSiC Project, 2009
16. Jurčíček, F., Gašić, M., Keizer, S., Mairesse, F., Thomson, B., Yu, K., Young, S.: Transformation-based learning for semantic parsing. In: Proceedings of Interspeech, pp. 2719–2722. ISCA, 2009
17. Kate, R.J., Wong, Y.W., Mooney, R.J.: Learning to transform natural to formal languages. In: Proceedings of AAAI, 2005
18. Kate, R.J.: A dependency-based word subsequence kernel. In: Proceedings of EMNLP, 2008
19. Mairesse, F.: Training tools and semantic decoder for the towninfo task: D2.1 (prototype). Technical Report D2.1, CLASSiC, February 2009
20. Mairesse, F., Gašić, M., Jurčíček, F., Keizer, S., Thomson, B., Yu, K., Young, S.: Spoken language understanding from unaligned data using discriminative classification models. In: Proceedings of ICASSP, 2009
21. Marcus, M.P., Santorini, B., Marcinkiewicz, M.A.: Building a large annotated corpus of English: The Penn Treebank. Computational Linguistics **19**(2), 313–330 (1993)
22. Merlo, P., Musillo, G.: Semantic parsing for high-precision semantic role labelling. In: Proceedings of the 20th Conference on Computational Natural Language Learning (CoNLL 2008), Manchester, UK (2008)
23. Meza-Ruiz, I.V., Riedel, S., Lemon, O.: Spoken Language Understanding in dialogue systems, using a 2-layer Markov Logic Network: Improving semantic accuracy. In: Proceedings of Londial, 2008
24. Palmer, M., Gildea, D., Kingsbury, P.: The proposition bank: An annotated corpus of semantic roles. Computational Linguistics **31**(1), 71–106 (2005)
25. van der Plas, L., Henderson, J., Merlo, P.: Domain adaptation with artificial data for semantic parsing of speech. In: Proceedings of the 2009 Annual Conference of the North American Chapter of the Association for Computational Linguistics, Companion Volume: Short Papers, pp. 125–128. Boulder, Colorado, June 2009
26. Ruppenhofer, J., Ellsworth, M., Petruck, M., Johnson, C., Scheffczyk, J.: Framenet ii: Extended theory and practice. Technical report, Berkeley, CA (2010)
27. Surdeanu, M., Johansson, R., Meyers, A., Màrquez, L., Nivre, J.: The CoNLL-2008 shared task on joint parsing of syntactic and semantic dependencies. In: Proceedings of the 12th Conference on Computational Natural Language Learning (CoNLL-2008), 2008
28. Thomson, B., Yu, K., Gašić, M., Keizer, S., Mairesse, F., Schatzmann, J., Young, S.: Evaluating semantic-level confidence scores with multiple hypotheses. In: Proceedings of the Ninth Conference of the International Speech Communication Association (INTERSPEECH 2008), pp. 1153–1156. Brisbane, Australia (2008)
29. Thomson, B., Gašić, M., Keizer, S., Mairesse, F., Schatzmann, J., Yu, K., Young, S.: User study of the Bayesian update of dialogue state approach to dialogue management. In: Proceedings of Interspeech, 2008
30. Ward, W.: Understanding spontaneous speech: the phoenix system. In: Proceedings of ICASSP, vol. 1, pp. 365–367, 1991
31. Williams, J.: Applying POMDPs to Dialog Systems in the Troubleshooting Domain. In: Proceedings of HLT/NAACL Workshop on Bridging the Gap: Academic and Industrial Research in Dialog Technology
32. Wu, W.-L., Lu, R.-Z., Duan, J.-Y., Liu, H., Gao, F., Chen, Y.-Q.: Spoken language understanding using weakly supervised learning. Computer Speech & Language, **24**(2), 358–382 (2010)
33. Young, S.: CUED standard dialogue acts. Technical report, Cambridge University Engineering Dept., 2007
34. Zettlemoyer, L.S., Collins, M.: Online learning of relaxed CCG grammars for parsing to logical form. In: Proceedings of EMNLP-CoNLL, 2007
35. Zhou, D., He, Y.: Discriminative Training of the Hidden Vector State Model for Semantic Parsing. IEEE Transactions on Knowledge and Data Engineering **21**(1), 66–77 (2009)

Chapter 4
User Simulation in the Development of Statistical Spoken Dialogue Systems

Simon Keizer, Stéphane Rossignol, Senthilkumar Chandramohan, and Olivier Pietquin

Statistical approaches to dialogue management have steadily increased in popularity over the last decade. Recent evaluations of such dialogue managers have shown their feasibility for sizeable domains and their advantage in terms of increased robustness. Moreover, simulated users have shown to be highly beneficial in the development and testing of dialogue managers and in particular, for training statistical dialogue managers. Learning the optimal policy of a POMDP dialogue manager is typically done using the reinforcement learning (RL), but with the RL algorithms that are commonly used today, this process still relies on the use of a simulated user. Data-driven approaches to user simulation have been developed to train dialogue managers on more realistic user behaviour. This chapter provides an overview of user simulation techniques and evaluation methodologies. In particular, recent developments in agenda-based user simulation, dynamic Bayesian network-based simulations and inverse reinforcement learning-based user simulations are discussed in detail. Finally, we will discuss ongoing work and future challenges for user simulation.

4.1 Introduction

Ever since their first introduction more than a decade ago, now, statistical approaches to dialogue management [24, 42] have grown in popularity. Evaluations of such spoken dialogue systems have shown their feasibility and benefits in various

S. Keizer (✉)
Heriot-Watt University, Edinburgh, EH14 4AS, UK
e-mail: s.keizer@hw.ac.uk

S. Rossignol • S. Chandramohan • O. Pietquin
SUPELEC, Supélec Campus de Metz, 2 rue Edouard Belin, 57070 Metz, France
e-mail: stephane.rossignol@supelec.fr; senthilkumar.chandramohan@supelec.fr; olivier.pietquin@supelec.fr

O. Lemon and O. Pietquin (eds.), *Data-Driven Methods for Adaptive Spoken Dialogue Systems*, DOI 10.1007/978-1-4614-4803-7_4,

domains. In particular, the probabilistic framework of Partially Observable Markov decision processes (POMDPs) has resulted in systems that are more robust in noisy conditions. Generally, state-of-the-art dialogue managers have two main tasks: (1) maintaining an information state throughout a dialogue, and (2) deciding on appropriate responses given that information state. In a POMDP dialogue manager, the former task is reformulated as *belief monitoring*, where, instead of a single information state, a probability distribution over possible information states (the *belief state*) is maintained. The action selection task consists of executing a dialogue *policy* which maps belief states to response actions, usually via some form of *value function*. The parameters for the belief monitoring component can be learned directly from data, which typically consists of estimating probabilities that govern a so-called *user model*, which relates the observed user behaviour to the user's state and more specifically, the user's goal. Learning an optimal policy is typically done using reinforcement learning (RL). For large-scale real-world domains, however, online policy optimisation from conversations with real users requires too many dialogues to be feasible and effective given the RL techniques currently used[1], and therefore relies on the availability of user simulations. It is important to distinguish between a user *simulation* which is an independent module by itself and a user *model* which is part of the dialogue manager. Although one could use certain user models to generate user behaviour as well, this behaviour is generally not realistic and complex enough to be used for policy optimisation.

In a statistical framework for dialogue management where policy optimisation is performed in interaction with a simulated user, the quality of the resulting dialogue system relies heavily on the quality of the user simulation. In order to get positive results from a system evaluation on real users, it is therefore important for the simulator to generate realistic user behaviour. In the learning phase, the system will then explore those parts of the state space which it will likely encounter in interaction with real users. For this reason, learning user simulation models from data is an important direction of research in this field [9, 13, 41].

In this chapter we will review several approaches to user simulation and discuss methodologies for evaluating simulated users. In particular, we will discuss recent developments regarding the *agenda-based* approach to user simulation [40], the approach of *dynamic Bayesian network* based simulation, and *inverse reinforcement learning* (IRL) based user simulation. The chapter will be concluded with a discussion of the methods presented as well as ongoing and planned work dealing with future challenges.

4.2 User Simulation Techniques: An Overview

Over the years, a variety of user simulation techniques have been developed to support the development of spoken dialogue systems. User simulation models that

[1]However, recent work on fast policy optimisation for POMDP dialogue managers (see Chap. 5) has shown promising results in online learning with human users [12], and another promising trend of research involves directly learning a policy from data [27].

can be trained offline from a dialogue corpus have proven to be very useful for testing a dialogue system before initiating the costly and time-consuming effort of evaluating the system with human subjects. Although one could think of potential advantages to acoustic-level [14] and word-level simulations [16], we will focus here on the more robust, portable and scalable *intention-level* user simulations. Such simulations take system behaviour as input and produce user behaviour as output, both semantically specified in the form of *dialogue acts*.

One of the first data-driven simulations was introduced for the purpose of testing and evaluation and consisted of a relatively simple probabilistic model where user actions are generated from a conditional probability distribution over user actions, given the previous system action [11]. This dialogue act bigram user simulator has also been used for learning dialogue strategies in a Markov decision process (MDP) framework for dialogue management [24]. In order to account for the overall consistency of user behaviour during a dialogue, [41] developed a probabilistic model conditioned on global user states instead of only the last system action. For training, a user goal inference model was used to provide the necessary data for maximum likelihood estimation of the probabilities. A similar approach was taken by [13] who developed a simulation model for training information state update (ISU) dialogue managers. The data required for training the user simulator was acquired via a process of semi-automatic annotation of dialogues in a corpus. As an alternative, unsupervised learning approaches to training user simulation models were also developed, which only require data on annotated user and system actions. Cuayáhuitl et al. [9] developed a model based on HMMs, whereas [39] presented an unsupervised learning method for training an agenda-based simulator.

In Sect. 4.4, we will describe a more recently developed method for parameter estimation in an agenda-based user simulator. The method combines a user goal inference model with sampling, allowing supervised learning of the parameters that control stochastic decisions in the simulator.

Another data-driven, probabilistic approach to user simulation that goes beyond n-grams is the use of dynamic Bayesian networks. Using evidence based on the input system behaviour, posterior distributions can be computed for the user output behaviour, using standard probabilistic inference techniques [30]. In Sect. 4.5, an overview will be given of recent work in this area.

In the next section, we discuss the various ways of evaluating user simulations. For a more detailed and complete overview of user simulation techniques, see [38].

4.3 Evaluation of Simulated Users

The quality of simulated users for developing, testing and training dialogue managers is of crucial importance because it dramatically influences the performance of the dialogue system when evaluated on human users. The assessment of the quality of simulated dialogues and user simulation methods is still an open issue, and there is no single commonly adopted evaluation metric. There are, however, various desired properties that a simulated user should have, and all metrics developed so far take these properties into account to some extent, depending partly on the purpose of the simulation.

First of all, a simulator should not only generate behaviour that matches the statistics of human user behaviour (as found in a dialogue corpus), but also generalise well to produce realistic unseen user behaviour. Secondly, the performance of a dialogue system in interaction with a simulated user should accurately predict its performance in interaction with real users. Thirdly, when using a simulator for training a dialogue manager, a good simulator should result in dialogue management strategies that perform well in interaction with real users. Hence, a suitable evaluation metric should assess these user simulation properties, and can then be used to rank simulation models as well as serve as an optimisation criterion for training dialogue managers.

In [38], two main categories of evaluation metrics are distinguished: (1) direct methods which assess the quality of predictions made by a simulator and (2) indirect methods which assess the performance of strategies learned with a simulator. Here, we follow the distinction made by [31] between (1) local methods which measure turn-level statistics and (2) global methods which measure dialogue-level statistics. Some work has been done on human evaluation of user simulations which can serve as a gold standard [2], but we focus here on automatic evaluation.

4.3.1 *Turn-Level Metrics*

Most of the early proposed metrics measure local consistency of simulated data with data from real users at turn level. Besides comparing full dialogue act statistics of simulated and real user behaviour in general, one can also address more specific aspects such as dialogue style (by comparing the relative frequencies of dialogue act types, e.g., confirm or question) or user cooperativeness (by comparing the proportion of slot values provided when requested).

4.3.1.1 Precision, Recall, and (Expected) Accuracy

Precision and Recall are common measures in machine learning and information retrieval and measure how well a model predicts observed values. A user model can be considered as a predictor of a dialogue act given some context [46]. Similar metrics, adapted from user modelling literature [38], are widely used in user simulation and even outside the realm of spoken dialogue systems. Precision and recall are here defined as

$$\text{Precision:}\; P = 100 \times \frac{\text{Correctly predicted actions}}{\text{All actions in simulated response}}. \tag{4.1}$$

$$\text{Recall: } R = 100 \times \frac{\text{Correctly predicted actions}}{\text{All actions in real response}}. \tag{4.2}$$

These two measures are complementary and cannot be used individually to rank user simulation methods. However, the balanced F-measure [34] can be used to combine these measures into a single scalar:

$$F = \frac{2PR}{P+R}. \tag{4.3}$$

Other related metrics are accuracy and expected accuracy as firstly introduced by [46] and adapted by [13]:

- Accuracy: "percentage of times the event that actually occurred was predicted with the highest probability"
- Expected accuracy: "Average of the probabilities with which the event that actually occurred was predicted"

One of the major drawbacks of these metrics is that they do not measure the generalisation capabilities of the user simulation. In fact, these metrics actually penalise attempts to generalise since when the model generates unseen dialogues, their scores are lower.

4.3.1.2 Kullback-Leibler Divergence and Dissimilarity

An alternative metric for comparing turn-level dialogue act statistics is the Kullback-Leibler (KL) divergence [19], which can be used to compare distributions of dialogue acts. It was introduced as an evaluation metric for user simulation in [9], where an HMM-based generative model for user simulation was proposed. The KL divergence between two distributions P and Q is defined by:

$$D_{\mathrm{KL}}(P||Q) = \sum_{i=1}^{M} p_i \log\left(\frac{p_i}{q_i}\right), \tag{4.4}$$

where p_i (resp. q_i) is the frequency of dialogue act a_i in the histogram of distribution P (resp. Q). Actually, the KL divergence is not a distance since it is not symmetric ($D_{\mathrm{KL}}(P||Q) \neq D_{\mathrm{KL}}(Q||P)$). To remedy this defect, the dissimilarity metric $DS(P||Q)$ is introduced:

$$DS(P||Q) = \frac{D_{\mathrm{KL}}(P||Q) + D_{\mathrm{KL}}(Q||P)}{2}. \tag{4.5}$$

Clearly, the KL divergence better captures the similarity between turn-level dialogue act statistics of simulated and real user behaviour than the precision and recall metric. However, it is an unbounded metric which makes it difficult to use for ranking. In addition, it gives more importance to the similarity of the means of the two distributions than to the similarity of the variances. Therefore, two distributions having the same means but very different variances will appear to be closer to each other than two distributions having slightly different means but similar variances.

KL divergence also requires a correct estimation of densities P and Q whilst traditionally, only counts are available from data. It is also difficult to assess the generalisation capabilities of a user model with this metric since it penalises dialogue strategies which are different from the real data.

4.3.2 Dialogue-Level Metrics

In this section, metrics are presented that are based on properties of entire dialogues instead of turns in a limited local context. For the turn-level metrics precision/recall and KL divergence, generalisations have been proposed to measure similarities between sequences of dialogue acts.

4.3.2.1 Perplexity and Log-Likelihood

Perplexity is a measure that comes from information theory and was first proposed as a user simulation metric in [13]. It is generally used to compare probabilistic predictive models. In natural language processing, it is widely used to compare language models. The perplexity of a model is defined as follows:

$$PP = 2^{\sum_{i=0}^{N} \frac{1}{N} \log_2 p_m(x_i)}, \tag{4.6}$$

where $p_m(x_i)$ is the probability of x_i given the model and x_i is a sample from a test dataset containing N samples. If the model is good, it will tend to give high probabilities to the test samples since it is supposed to predict them and, therefore, have a low perplexity. In the case of a user model, the data to be predicted are sequences of dialogue acts $\{a_0, a_1, \ldots, a_n\}$, so $p_m(x) = p_m(a_0, a_1, \ldots, a_n)$ is the probability of a sequence of dialogue acts given the user model.

A similar metric is the log-likelihood $\mathcal{L}(x)$ of a data set $x = \{x_i\}_{i=1,\ldots,N}$ given a model m, defined by $\mathcal{L}(x) = \log p(x|m) = \log p_m(x)$. If the data samples x_i are assumed to be independent (a common assumption), $\mathcal{L}(x)$ can be written as:

$$\mathcal{L}(x) = \log \prod_{i=1}^{N} p_m(x_i) = \sum_{i=1}^{N} \log p_m(x_i). \tag{4.7}$$

The higher the log-likelihood, the higher the consistency between the data and the model.

Perplexity and log-likelihood measure how well a model is able to predict data in a held-out test set and therefore can be used to measure generalisation. Perplexity and log-likelihood are scalar numbers that can be used to rank user simulations; however, the perplexity figure can be difficult to interpret as it ranges from 1 to infinity. These metrics can be viewed as global or dialogue-level metrics as they are based on the probability of sequences of dialogue acts. However, it is somewhat difficult to assign relative importance of dialogue-level features such as dialogue length and task completion.

4.3.2.2 BLEU and Discourse-BLEU

The BLEU (Bilingual Evaluation Understudy) score [26] is widely used in machine translation. It is a metric that compares two semantically equivalent sentences by computing the geometric mean of the n-gram precisions with a brevity penalty to compensate for high n-gram precision of short utterances. This metric can also be used to compare user simulations that generate word-level output, as proposed by [16]. The BLEU metric can thus be used to measure the naturalness of a simulated utterance, but what we are interested in is a measure of naturalness of dialogues (i.e., sequences of utterances). To achieve this goal, [16] propose another metric they call Discourse-BLEU (D-BLEU). D-BLEU is also the geometric mean of the n-gram precisions with a brevity penalty, but the n-grams considered here are not sequences of words but sequences of intentions.

The BLEU score is known to be highly correlated with human judgment [10, 26], and [16] argue that D-BLEU also follows the same tendencies as human judgment. In some cases, BLEU has been reported to fail to predict translation improvements and naturalness [21]. Too few studies are reported about the D-BLEU score for dialogue to allow one to draw the same conclusions, but the BLEU and D-BLEU are quite similar in their definitions. Once again, this metric also fails to measure the generalisation capabilities of the user simulation.

4.3.2.3 SUPER

SUPER (Simulated User Pragmatic Error Rate) [32, 33] is an attempt to combine different metrics so as to take advantage of their respective features. Similar to computing word error rate for speech recognition systems in terms of insertions, deletions and substitutions, this metric combines scores for consistency (in terms of insertions of intentions), completeness (in terms of deletions of intentions) and variety (comparing probabilities of human and simulated user intentions). For each context k, these scores I_m, D_m and V_m, respectively, are computed and combined into a single score as follows:

$$\text{SUPER} = \frac{1}{m}\sum_{k=1}^{m}\frac{V_m + I_m + D_m}{n}, \tag{4.8}$$

where n is the number of possible user intentions and m the number of contexts.

The SUPER score has many of the desired features of an evaluation metric for user simulation, but it is not a direct measure of the ability to predict the performances of a spoken dialogue system when used with real users.

4.3.2.4 Predicted System Performance

When evaluating a dialogue system in interaction with a user simulator, the results should accurately predict the performance of the system in interaction with real users. A user simulator can therefore be evaluated by measuring the similarity between performance scores obtained from a simulation-based evaluation and those obtained from a human-based evaluation of the same system. System performance is generally measured in terms of task success rate or an average dialogue score, where a single dialogue score takes into account several factors including task completion, the number of turns taken and possibly other factors considered relevant (see also Chap. 7).

A more advanced way of measuring the predictive capability of a user simulator in terms of system performance is to look at distributions over dialogue scores. Williams [44] proposes to use the normalised Cramér-von Mises (CvM) divergence [3, 8] to measure the similarity between such distributions. The CvM divergence provides a real number ranging from 0 to 1 and is computed as follows:

$$D_{\text{CvM}}(F_0||F_1) = \alpha \sqrt{\sum_{i=1}^{N_0} \left(F_0\left(x^0_{(i)}\right) - F_1\left(x^0_{(i)}\right)\right)^2}, \tag{4.9}$$

where F_j is the empirical distribution function (EDF) of the data $S_j = \left(x^j_{(1)}, x^j_{(N_j)}\right)$ and α is a normalising constant given by $\alpha = \sqrt{\frac{12N_0}{4N_0^2-1}}$. By definition, the EDF is:

$$F_j(x) = \frac{1}{N_j} \sum_{i=1}^{N_j} \begin{cases} 1 & \text{if } x^j_{(i)} < x \\ \frac{1}{2} & \text{if } x^j_{(i)} = x \\ 0 & \text{if } x^j_{(i)} > x. \end{cases} \tag{4.10}$$

Being based on the *EDF*, the CvM metric does not make any assumptions about the underlying distributions. This is an advantage over the KL-divergence discussed above, which is accurate only if the underlying distributions are known or well approximated. Also, the CvM divergence gives a bounded scalar value, making it suitable for ranking.

4.3.2.5 Performance of Learnt Strategies

The ultimate test for a user simulator when used for training RL-based dialogue managers is the evaluation with real users of strategies that are learned in interaction with that simulator. Such an evaluation will tell us what is the effect of a simulator on system performance. Hence, a given dialogue manager can be trained on different user simulators, resulting in different dialogue strategies that can be evaluated on human users. The simulators can then be ranked according to the performance of the corresponding strategies.

4.3.2.6 Inverse Reinforcement Learning Metric

Recently, a new approach to learning user simulation models has been developed, which is based on IRL [25]. In this approach [6], which will be explained in more detail in Sect. 4.6, user behaviour as observed in a corpus of human–machine dialogues is considered to be optimal, and the idea is to discover a reward function that could have resulted in that optimal behaviour. Once this reward function is learnt, an artificial agent can be trained using this function so as to mimic the expert's behaviour. This inferred reward function can also be regarded as an evaluation metric for user simulation. A higher cumulative reward indicates that the user behaviour is more similar to the expert user behaviour found in the corpus it was inferred from. Such a metric has the advantage of not being sensitive to a simulator generating unseen dialogues and therefore better captures the generalisation capability of a simulated user. Disadvantages include the metric's sensitivity to specific user types and its potential inaccuracy in case a user's behaviour in the corpus is not optimal with respect to their reward function.

4.4 Agenda-Based User Simulation

In agenda-based user simulation, user acts are generated on the basis of a *user goal* and an *agenda* [40]. The simulator presented here is developed and used for a tourist information application but is sufficiently generic to accommodate slot-filling applications in any domain[2]. The user goal consists of the type of venue, for example, `hotel`, `bar` or `restaurant`, a list of constraints in the form of slot-value pairs, such as `food=Italian` or `area=east`, and a list of slots the user wants to know the value of, such as the address (`addr`), phone number (`phone`) or price information (`price`) of the venue. The user goals for the simulator are randomly generated from the domain ontology describing which combinations of venue types and constraints are allowed and what are the possible values for each slot. The agenda is a stack-like structure containing planned user acts. When the simulator receives a system act, the status of the user goal is updated as well as the agenda, typically by pushing new acts onto it. In a separate step, the response user act is selected by popping one or more items off the agenda.

---

[2] We have to date also implemented systems in appointment scheduling and bus timetable inquiries.

Although the agenda-based user simulator introduced by [40] was entirely handcrafted, it was realistic enough to successfully test a prototype POMDP dialogue manager and train a dialogue policy that outperformed a handcrafted baseline [45]. A method to train an agenda-based user simulator from data was proposed by [39]. In this approach, operations on the agenda are controlled by probabilities learned from data using a variation of the EM algorithm. However, this approach does not readily scale to more complex interactions in which users can, for example, change their goal midway through a dialogue.

4.4.1 Random Decision Parameters

Each time the user simulator receives a system act, a complex, twofold process takes place involving several decisions, made on the basis of both the nature of the incoming system act and the information state of the user, i.e., the status of the user goal and agenda. The first phase can be seen as an ISU and involves actions like filling requested slots or checking whether the provided information is consistent with the user goal constraints. In the second phase, the user decides which response act to generate, based on the updated agenda. Many of the decisions involved are deterministic, allowing only one possible option given the context. Other decisions allow for some degree of variation in the user behaviour and are governed by probability distributions over the options allowed in that context. For example, if the system has offered a venue that matches the user's goal, the user can decide to either change their goal or to accept the venue and ask for additional information such as the phone number.

The non-deterministic part of the simulator is formalised in terms of a set of *random decision points* (RDPs) embedded in the decision process. If an RDP i is encountered (depending on the context), a random (i.e., stochastic) choice between the options defined for that point is made by sampling from a probability distribution, specified by parameters θ_i. Most of the RDPs are controlled by a multinomial distribution, such as deciding whether or not to change the goal after a system offer. The other RDPs are controlled by geometric distributions, for example, for deciding how many dialogue acts to pop from the agenda and combine into a single user response act. The parameter for this distribution specifies the probability of deciding to include no more planned dialogue acts into a single user response act.

Hence, the user simulator can be viewed as a "decision network," consisting of deterministic and random (stochastic) decision points. This is illustrated in Fig. 4.1 for the simplified case of a network with only four RDPs; the actual simulator has 23 RDPs, with 27 associated parameters in total. Each time the simulator receives a system act, it follows a path through the network, determined by that system act and the user goal and agenda, as well as by decisions made at any RDPs encountered.

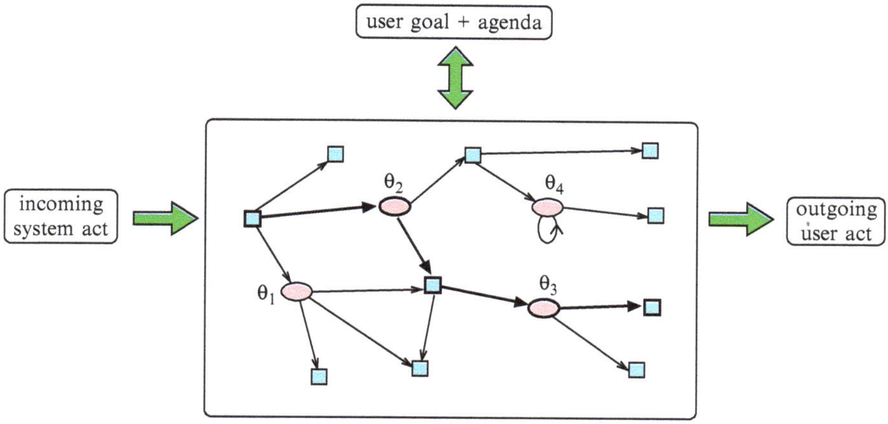

Fig. 4.1 User simulator viewed as a "decision network": square nodes indicate deterministic decision points; round nodes indicate random decision points and have associated parameters θ_i; the loop on one of the nodes indicates it has a geometric distribution associated with it

4.4.2 Training the Simulator from Data

The parameterisation of the user simulator as described in Sect. 4.4.1 forms the basis for a method for training the simulator with real user data. The parameters describing the probability distributions for each RDP are estimated in order to generate user behaviour that fits the user behaviour in the corpus as closely as possible. Because the random decisions underlying observed user acts cannot directly be inferred from the data, straightforward maximum likelihood estimation [41] is not possible. Therefore, a sampling approach is taken in which the simulator is run repeatedly against the system acts in the corpus in order to find the stochastic decisions that lead to simulated user acts matching the observed user acts. Once all combinations of decisions for the matching user acts are found, the counts can be used to make maximum likelihood estimates for the parameters.

4.4.2.1 Parameter Estimation

Before starting the process of matching simulated acts with true acts and collecting counts for the RDPs, the parameters are initialised to values corresponding to uniform distributions. Then, the simulator is run against all dialogues in the corpus in such a way that for each turn in a dialogue (consisting of a system act and a user act), the user simulator is provided with the system act and is run repeatedly to generate several simulated user response acts for that turn. For the first turn of a dialogue, the simulator is initialised with the correct user state (see Sect. 4.4.2.2). For each response, the simulator may make different decisions, generally leading to different user acts. The decisions that lead to a simulated act that matches the true act are recorded as successful. By generating a sufficiently large number of simulated acts, all possible combinations of decisions are explored to find a matching act. Given the high complexity of the simulator, this sampling approach is preferred over directly enumerating all decision combinations to identify the successful ones. If none of the combinations are successful, then either (a) the processing of the dialogue is ended or (b) the correct context is set for the next turn and processing is continued. Whereas the former approach aims at matching sequences of turns, the latter only aims at matching each user turn separately.

4.4.2.2 User Goal Inference

In order to be able to set the correct user goal state in any given turn, a set of update rules is used to infer the user's goals from a dialogue beforehand, on the basis of the entire sequence of system acts and "true" user acts (see Sect. 4.4.3.1) in the corpus. These update rules are based on the notion of *dialogue act preconditions*, which specify conditions of the dialogue context that must hold for a dialogue agent to perform that act. For example, a precondition for the act `inform(area=central)` is that the speaker wants a venue in the centre. The user act model of the HIS dialogue manager is designed according to this same notion [17]. For the user goal inference model, the user act is given, and therefore, its preconditions can be used to directly infer the user goal. So, for example, in the case of observing the user act `inform(area=central)`, the constraint `(area=central)` is added to the user goal.

In using this offline goal inference model, our approach takes a position between [39], in which the user's goal is treated as hidden, and [13], in which the user's goal is obtained directly from the corpus annotation.

4.4.3 Evaluation

Following the distinction between *direct* and *indirect* evaluation methods made in [38], two different evaluation methods were applied for evaluating the parameter estimation technique for training the user simulator. The first evaluation involved comparing the statistics of simulated and real user behaviour, whereas the second evaluation involved comparing dialogue manager policies trained with different simulators.

4.4.3.1 Data

The dialogue corpus that was used for training and evaluating the simulator was obtained from the evaluation of a POMDP spoken dialogue system with real users in the tourist information domain. All user utterances in the resulting corpus were transcribed and semantically annotated in terms of dialogue acts. Dialogue acts consist of a series of semantic items, including the type (describing the intention of the speaker, e.g. `inform` or `request`) and a list of slot value pairs (e.g. `food=Chinese` or `area=south`). An extensive analysis of the annotations from three different people revealed a high level of inter-annotator agreement (ranging from 0.81 to 0.94, depending on which pair of annotations are compared), and a voting scheme for selecting a single annotation for each turn ensured the reliability of the "true" user acts used for training the simulator [18].

4.4.3.2 Corpus Statistics Results

A first approach to evaluating user simulations is to look at the statistics of the user behaviour that is generated by a simulator and compare it with that of real users as observed in a dialogue corpus. From the metrics outlined in Sect. 4.3, we computed the log-likelihood of the data given the user simulation model, the turn-level precision and recall, and the KL divergence between simulated and real user act distributions. The computation of these metrics for the agenda-based simulator with a particular set of parameters is not straightforward and requires a sampling setup similar to the one used for training; see [18] for more details. Also, it should be noted that the log-likelihood only represents those turns in the corpus for which the simulated user can produce a matching simulated act with some probability. Hence, it is important to also take into account the *corpus coverage* when considering the log-likelihood in corpus-based evaluation. Dividing by the number of matched turns provides a useful normalisation in this respect.

For the experiments, the corpus data was randomly split into a training set, consisting of 4,479 user turns in 541 dialogues, used for estimating the user simulator parameters, and a test set, consisting of 1,457 user turns in 175 dialogues, used for evaluation only. In the evaluation, the following parameter settings were compared: (1) non-informative, uniform parameters (UNIF), (2) handcrafted parameters (HDC), (3) parameters estimated from data (TRA), and (4) deterministic parameters (DET), in which for each RDP, the probability of the most probable decision according to the estimated parameters is set to 1, i.e. at all times, the most likely decision according to the estimated parameters is chosen.

For both trained and deterministic parameters, a distinction is made between the two approaches to matching user acts during parameter estimation. Recall that in the turn-based approach, in each turn, the simulator is run with the corrected context to find a matching simulated act, whereas in the sequence-based approach, the matching process for a dialogue is stopped in case a turn is encountered which cannot be matched by the simulator. This results in estimated parameters TRA-T and deterministic parameters DET-T for the turn-based approach and analogously TRA-S and DET-S for the sequence-based approach. The corresponding normalised log-likelihoods are indicated by nLL-T and nLL-S.

Tables 4.1 and 4.2 give the results on the training and test data, respectively. The results show that in terms of log-likelihood and KL dissimilarity (DS-Full for distributions over dialogue acts including their semantic content and DS-Types for distributions over dialogue act types only), the estimated parameters outperform the

Table 4.1 Results of the sample-based user simulator evaluation on the Mar'09 training corpus (the corpus coverage was 59 % for the turn-based and 33 % for the sequence-based matching approach)

PAR	nLL-Turn	nLL-Seq	Precision	Recall	F-Score	DS-Full	DS-Types
UNIF	−3.78	−3.37	16.95 (±0.75)	9.47 (±0.59)	12.15	3.057	2.318
HDC	−4.07	−2.22	44.31 (±0.99)	34.74 (±0.95)	38.94	1.784	0.623
TRA-T	**−2.97**	–	37.60 (±0.97)	28.14 (±0.90)	32.19	1.362	0.336
DET-T	−∞	–	47.70 (±1.00)	40.90 (±0.98)	44.04	2.335	0.838
TRA-S	–	**−2.13**	43.19 (±0.99)	35.68 (±0.96)	39.07	**1.355**	**0.155**
DET-S	–	−∞	**49.39** (±1.00)	**43.04** (±0.99)	**46.00**	2.310	0.825

Table 4.2 Results of the sample-based user simulator evaluation on the Mar'09 test corpus (corpus coverage 59 % for the turn-based and 36 % for sequence-based matching)

PAR	nLL-Turn	nLL-Seq	Precision	Recall	F-Score	DS-Full	DS-Types
UNIF	−3.61	−3.28	16.59 (±1.29)	9.32 (±1.01)	11.93	2.951	2.180
HDC	−3.90	−2.19	45.35 (±1.72)	36.04 (±1.66)	40.16	1.780	0.561
TRA-T	**−2.84**	–	38.22 (±1.68)	28.74 (±1.57)	32.81	**1.405**	0.310
DET-T	−∞	–	49.15 (±1.73)	42.17 (±1.71)	45.39	2.478	0.867
TRA-S	–	**−2.12**	43.90 (±1.72)	36.52 (±1.67)	39.87	1.424	**0.153**
DET-S	–	−∞	**50.73** (±1.73)	**44.41** (±1.72)	**47.36**	2.407	0.841

other settings, regardless of the matching method. In terms of precision/recall (given in percentages with 95 % confidence intervals), the estimated parameters are worse than the handcrafted parameters for turn-based matching, but have similar scores for sequence-based matching.

The results for the deterministic parameters illustrate that much better precision/recall scores can be obtained, but at the expense of variability as well as the KL dissimilarity. It will be easier to train a dialogue policy on such a deterministic simulator, but that policy is likely to perform significantly worse on the more varied behaviour generated by the trained simulator, as we will see in Sect. 4.4.3.3.

Out of the two matching approaches, the sequence-based approach gives the best results: TRA-S outperforms TRA-T on all scores, except for the coverage which is much lower for the sequence-based approach (33 % vs. 59 %).

4.4.3.3 Policy Evaluation Results

Although the corpus-based evaluation results give a useful indication of how realistic the behaviour generated by a simulator is, what really should be evaluated is the dialogue management policy that is trained using that simulator. Therefore, different parameter sets for the simulator were used to train and evaluate different policies for the hidden information state (HIS) dialogue manager [45]. Four different policies were trained: one policy using handcrafted simulation parameters (POL-HDC), two policies using simulation parameters estimated (using the sequence-based matching approach) from two data sets that were obtained by randomly splitting the data into two parts of 358 dialogues each (POL-TRA1 and POL-TRA2) and finally, a policy

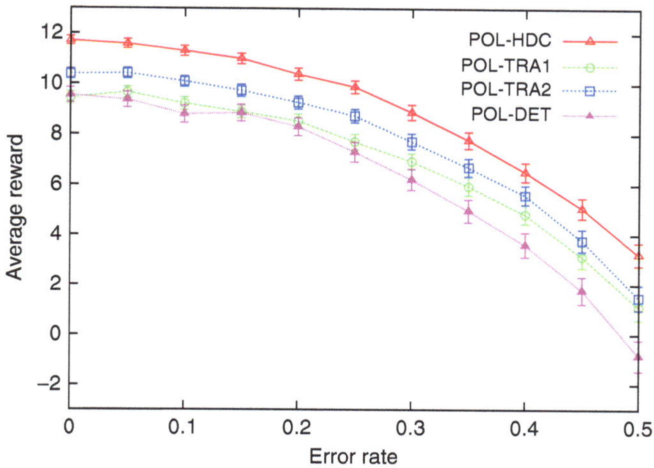

Fig. 4.2 Average rewards for each policy when evaluated on UM-HDC

using a deterministic simulator (POL-DET) constructed from the trained parameters as discussed in Sect. 4.4.3.2. The policies were then each evaluated on the simulator using the four parameter settings at different semantic error rates.

The performance of a policy is measured in terms of a reward that is given for each dialogue, i.e. a reward of 20 for a successful dialogue, minus the number of turns. A dialogue is considered successful if the system has offered a venue matching the predefined user goal constraints and has given the correct values of all requested slots for this venue. During the policy optimisation, in which a reinforcement learning algorithm tries to optimise the expected long term reward, this dialogue scoring regime was also used.

In Figs. 4.2–4.5, evaluation results are given resulting from running 3,000 dialogues at each of 11 different semantic error rates. The curves show average rewards with 95 % confidence intervals. The error rate is controlled by a handcrafted error model that converts the user act generated by the simulator into an n-best list of dialogue act hypotheses.

The policy that was trained using the handcrafted simulator (POL-HDC) outperforms the other policies when evaluated on that same simulator (see Fig. 4.2), and both policies trained using the trained simulators (POL-TRA1 and POL-TRA2) outperform the other policies when evaluated on either trained simulator (see Figs. 4.3 and 4.5). There is little difference in performance between policies POL-TRA1 and POL-TRA2, which can be explained by the fact that the two trained parameter settings are quite similar, in contrast to the handcrafted parameters. The policy that was trained on the deterministic parameters (POL-DET) is competitive with the other policies when evaluated on UM-DET (see Fig. 4.4), but performs significantly worse on the other parameter settings which generate the variation in behaviour that the dialogue manager did not encounter during training of POL-DET.

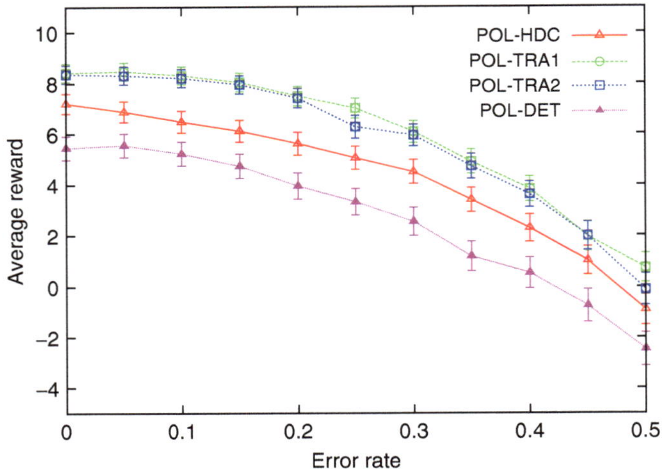

Fig. 4.3 Average rewards for each policy when evaluated on UM-TRA1

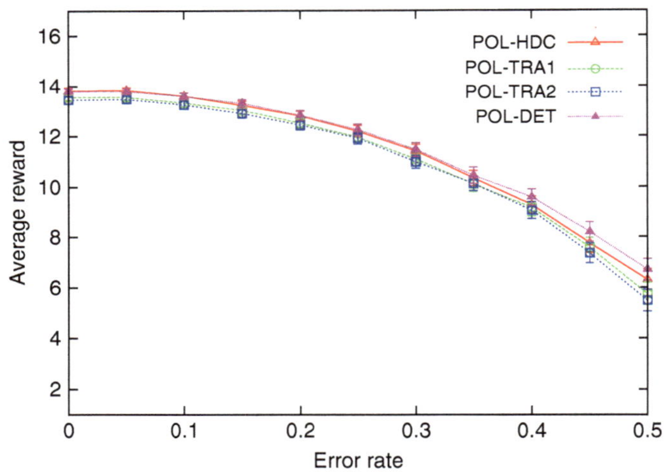

Fig. 4.4 Average rewards for each policy when evaluated on UM-DET

4.4.4 Summary of Agenda-Based User Simulation Parameter Learning

In this section we have presented a maximum likelihood approach combined with a sampling procedure for estimating the random decision parameters of the agenda-based user simulator. The results of the corpus statistics evaluation discussed in Sect. 4.4.3.2 show that the dialogues generated by the trained simulator

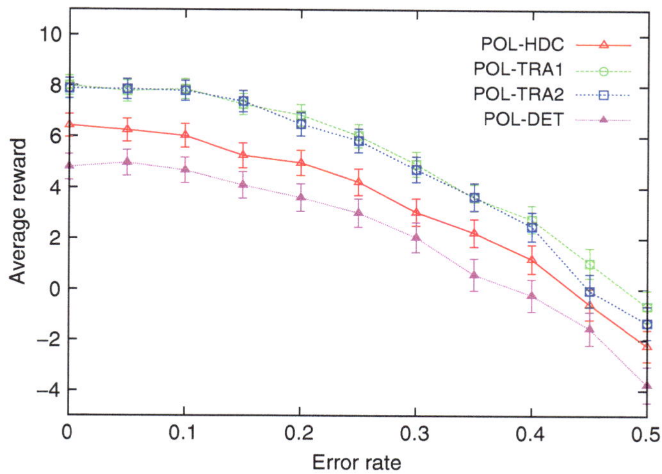

Fig. 4.5 Policy evaluation results in terms of average reward using the second trained simulator (UM-TRA2)

more closely match real corpus data. Furthermore, the results of the evaluation in terms of trained policy performance discussed in Sect. 4.4.3.3 show that the policies optimised with the trained user simulator are at least as robust as the other policies. Therefore, it seems likely that these trained policies will show improved performance when evaluated on real users.

However, this claim can only be properly demonstrated in a real user evaluation of the dialogue system running different dialogue management policies. Such a user trial would also be able to confirm whether the results from evaluations on the trained simulator can more accurately predict the actual performance expected with real users.

4.5 Dynamic Bayesian Network-Based User Simulation

Another user simulation approach in which user dialogue acts are generated on the basis of not only the last system action but also a richer notion of context than previously used is based on *Bayesian networks* (BNs). BNs are probabilistic models with a graphical structure representing conditional independence assumptions. Being generative models, they are not only suitable for inference, but also for generating data and can therefore be used for simulating user behaviour. BN-based user simulations provide a statistical framework for generating a wide range of synthetic dialogues that are statistically consistent with real data. The parameters of the BN user simulation parameters can be learned from a dialogue corpus or manually set using expert knowledge. The availability of numerous tools for BN training and inference ensures that BN based user simulations can be built with relative ease. The approach presented here is similar to the one proposed in [28, 30] but has three novel contributions: (i) the model is trained on a corpus of human–machine dialogues and tested on a state-of-the-art dialogue system, (ii) the trained model has been updated with expert knowledge to support grounding behaviour, and (iii) a new parameter estimation method has been developed for dealing with missing training data.

4.5.1 Description of the BN User Simulation

4.5.1.1 The Task

The dialogue task studied here is a slot-filling dialogue problem in the restaurant domain, similar to that in Sect. 4.4. The following slots are used: `Food` (`italian`, `indian` or `chinese`) `Pricerange` (`moderate`, `expensive` or `cheap`), and `Area` (`central`, `north`, `south`, `west` or `east`). The possible system dialogue act types are: `Hello`, `Request`, `Confirm`, `Close` and `ConfirmAndRequest`. The possible user dialogue act types are: `Inform`, `Close`, `Confirm`, `Negate` and `Null`.

4.5.1.2 Modelling User Simulation as a Bayesian Network

The interaction between the user and the dialogue manager (DM) is considered as a sequential transfer of intentions organised in dialogue turns. At each turn t, the DM selects a system act a_t, given its internal state s_t. The user responds with a user act u_t, based on user goal g_t and knowledge k_t about the dialogue progress (what information has been exchanged until reaching turn t). So at any given turn t, the information exchanged can be modelled as a joint probability $p(a,s,u,g,k)$ of all these variables:

$$p(a,s,u,g,k) = p(u|g,k,a,s)p(g|k,a,s)p(k|s,a)p(a|s)p(s). \tag{4.11}$$

Given that the user does not have access to the system's internal dialogue state s, the user act u, the user goal g and the user knowledge k cannot depend on s. Also, the user goal can only be modified on the basis of the user's knowledge of the dialogue. With these assumptions, $p(a,s,u,g,k)$ can be rewritten as

$$p(a,s,u,g,k) = \underbrace{p(u|g,k,a)}_{\text{User act}} \overbrace{p(g|k)}^{\text{Goal Modif.}} \underbrace{p(k|a)}_{\text{Know. update}} \overbrace{p(a|s)}^{\text{DM Policy}} p(s). \tag{4.12}$$

This can be expressed by the Bayesian network depicted in Fig. 4.6.

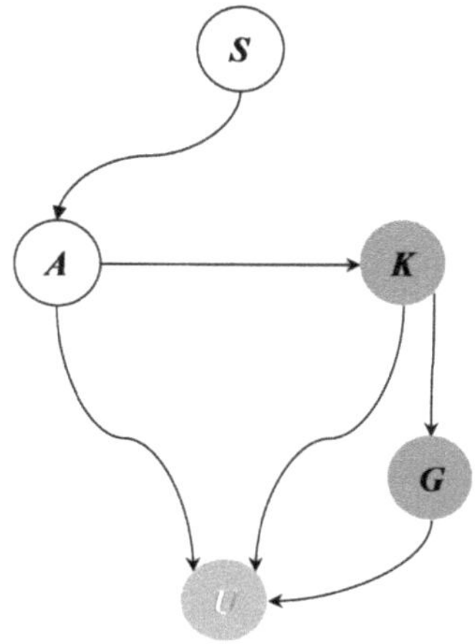

Fig. 4.6 Bayesian network-based user simulation

To use this kind of BN in practice, a tractable representation of the stochastic variables in the network is required. Therefore, the variable A is split into two sub-variables: the dialogue act type A_{Type} and the slot A_{Slot} to which it is applied. The user act U is also divided into two sub-variables: the user act type U_{Type} and the slot–value pairs U_{SVPs} to which they are applied (*e.g.*, U_{Type} = `inform`, U_{SVPs} = "`food=Italian`"). A new variable C is added to simulate the dialogue closure by the user simulation. The goal G is represented by three nodes, one for each slot, for example, $G_{Food} = Italian$, $G_{Pricerange} = cheap$, and $G_{Area} = dontcare$. These nodes are initialised with random values, and the simulator's behaviour will be consistent with that goal. The knowledge K is also represented by three nodes, but with possible values *low*, *medium* or *high* for each of the slots. These values indicate to what extent the user believes his preferred value for that slot has been transferred to the system. Both goal and knowledge nodes are internal variables, i.e., their value is unknown.

In the user simulation described in [24], the history of the dialogue and the goal of the user are not taken into account. Such a user simulation may suffer from lack of consistency for two reasons: (1) if the user simulation is prompted twice for the same slot, it can provide different values since the probabilities are not conditioned on a particular goal, and (2) if the user simulation is prompted infinitely for the same slot, it will keep giving the same answer infinitely since it does not "remember" that the slot has already been prompted. The introduction of KNW and GOAL nodes is an attempt to overcome this problem.

4.5.2 Training a BN-Based User Simulation

The process of learning the parameters of the BN-based user simulation is outlined in this section. The experimental results presented below suggest that BN user

Table 4.3 Comparison of trainedBN and heuristicBN user simulations

User type	Mean	Max	Min	4 turns	Less than 9 turns	More than 9 turns
heuristicBN	4.969	21	4	93.20 %	93.40	6.6 %
trainedBN	4.577	9	4	58.00 %	99.90	0.1 %

simulations with trained parameters generate synthetic dialogues that are more natural compared to BN user simulations using handcrafted parameters provided by experts. For the experiment, the dialogue corpus described in Sect. 4.4.3.1 was reused to train a three slot BN user simulation. In this section, it is assumed that the data required for training the BN parameters is complete, i.e., all variables are observed in the data. The BN parameters can thus be estimated using maximum likelihood, except for the knowledge parameters, which are fixed by an expert

4.5.2.1 Generalisation of the Training Task

Given the large number of BN parameters (2, 151 in total) and the small amount of data, it is not feasible to make good estimates for all probabilities. For example, the conditional probability distribution for variable C requires $5 \cdot 3^3 = 135$ probabilities, corresponding to the possible values of its parents A_{Type} and the knowledge nodes K_{Food}, $K_{Pricerange}$ and K_{Area}. Therefore, parameter tying is applied to reduce the number of parameters to determine. Instead of estimating all $3^3 = 27$ probabilities evolving from the knowledge nodes, only 3 probabilities are considered, and instead of the 5 probabilities evolving from the system act node, only 2 probabilities are considered, after assuming all system act types except "Close" to be equivalent. After additionally assuming that the system act node and the knowledge nodes are independent, only $2+3=5$ parameters remain to be estimated for variable C. Applying similar reasoning to the other nodes results in a set of only 25 probabilities to be either trained or heuristically fixed.

4.5.2.2 Experimental Setup and Evaluation Results

In order to quantify the effectiveness of the user simulation, the BN-based user simulation is made to interact with the REALL-DUDE /DIPPER dialogue management framework using the OAA environment (see [5, 7, 22]). The BN parameters can be trained from a dialogue corpus or heuristically fixed, resulting in two user simulations **trainedBN** and **heuristicBN**. For both user simulations, 1,000 dialogues were generated. Table 4.3 summarises the quality of these synthetic dialogues in terms of dialogue length.

It can be observed from Table 4.3 that the length of almost all dialogues generated using the heuristicBN user is 4 turns (93.2 %). This is due to the fact that the heuristicBN has been specifically designed to generate short dialogue episodes. However, some dialogues involve a large number of turns, which can be attributed to *grounding* problems, i.e. problems in reaching mutual understanding of the user's goal. For example, the user simulation sends *valueA* for *slot1*; the DM asks confirmation for *slot1*; the user sends *valueB* for *slot1*, simulating for instance ASR uncertainties; (at this stage, internally, the user simulation has a high value for the knowledge concerning *slot1*, and on the contrary, the DM has a very low value); when the other two slots are filled out, the user wants to stop the dialogue and sends a "Close" message rather than directly responding to the requests from the DM, causing the DM to get more and more confused. It is important to note that the REALL-DUDE/DIPPER/OAA environment prevents dialogues to be longer than 21 turns.

Statistical analysis of dialogue episodes generated using the trainedBN user shows that a large number of dialogue episodes are longer (compared to that of heuristicBN user dialogue). The percentage of dialogues which needed exactly 4 turns to accomplish the task drops from 93.2 % to 58.0 %. However, it can also be noticed that the mean number of turns for trainedBN is smaller when compared to heuristicBN. It can also be observed that very long dialogues (more than nine turns), which indicate some deep misunderstanding between the DM and the user simulation, have almost completely disappeared for the trainedBN. All these results indicate that the dialogues obtained with the trainedBN user are more natural, at least from the dialogue management point of view. In addition, user simulations based on the Bayesian network paradigm can efficiently interact with dialogue managers and thus generate synthetic dialogues that can be used for dialogue policy optimisation.

4.5.2.3 Coping with the Grounding Problem

If the DM asks confirmation for a slot for which the user simulation has a *low* knowledge value (i.e., a slot for which the user has never provided information yet), it is likely that a grounding problem has occurred. Using diagnostic inference in the Bayesian network of Fig. 4.6, one could infer the most probable dialogue state $\hat{s}_t$ at turn t given the observed DM act a_t and the user's knowledge k_t: $\hat{s}_t = \text{argmax}_s p(s|a_t, k_t)$. Recall that the user's knowledge keeps track of the dialogue progress from the user's point of view. Therefore, if the dialogue state estimate $\hat{s}_t$ is very different from the user knowledge k_t, then it is likely that a grounding problem has occurred. However, instead of computing the dialogue state estimate and comparing it to the user's knowledge, a Boolean valued node is added to the network, which explicitly represents the occurrence of a grounding problem [36]. If a grounding problem is detected for the slot i, the user simulation is forced to provide information concerning this slot.

Training was performed to learn the parameters of the updated BN user simulation. Using this *grounding-enabled* trainedBN user and the DIPPER dialogue management framework, 1,000 simulated dialogues are generated. In the generated dialogues, 496 grounding problems were detected by the simulator. Table 4.4 presents a dialogue example in which a grounding problem occurred and was

Table 4.4 Occurrence of grounding problem and BN user response

```
User goal: Food: italian, Price-range: cheap, Area: central
<syst act> hello(Food)
<user act> inform(slot_1='italian')
<syst act> confirm(Area) => grounding error detected for the slot_1
<user act> inform(slot_1='italian')
<syst act> request(rPrice)
<user act> inform(slot_2='cheap')
```

Table 4.5 Comparison of BN user simulation with and without grounding

User type	Mean no. of turns	More than 5 turns
No grounding BN user	5.254	29.22 %
Grounding BN user	5.069	20.29 %

detected and successfully dealt with by the updated BN user simulation. Considering that the dialogue task studied here involves three slots and since the user simulation is configured to send information about no more than one slot per turn, the minimum number of turns necessary to reach the end of the dialogue is four: one per slot and one "closingDialogue" turn. Statistics computed on dialogues with at least one grounding problem obtained with both the original trained BN user and the grounding-enabled BN user are presented in Table 4.5. It can be observed that the dialogues with the grounding-enabled BN user are considerably shorter. In fact, the number of dialogues longer than 5 turns is 30.6 % lower.

4.5.3 *Training an BN User Considering Data Missing from Corpus*

In using human–machine dialogue corpora for training, system designers increasingly have to deal with missing data. In this section, we discuss a method for training BN user simulations using algorithms which can deal with missing data. The focus here is on the user's internal representation of the dialogue context, referred to as the *knowledge* of the user. This is a major difference in comparison to past contributions such as [43] where transition probabilities are estimated according to the history of system and user acts. Taking into account the incremental knowledge of the user about previous exchanges is important to ensure the consistency of the dialogue during the interaction [29].

4.5.3.1 Experimental Setup

For training the models described in Sect. 4.5.2, it is assumed that the knowledge of the user can be inferred from the data itself and all other data required for training are available in the dialogue corpus. However, often the reality is that some aspects of training data may be missing. Therefore, we also experiment with training user simulations using incomplete data; in particular, the knowledge of the user is assumed to be hidden, and the BN parameters are learned using the expectation-maximisation (EM) algorithm, which is specifically designed to cope with missing data [35]. Training of the BN user simulation is conducted using six different configurations:

1. *Maximum Likelihood method* [**MaxL-BN**]: In this configuration, the data required for training the BN parameters is assumed to be complete. The knowledge parameters are fixed by an expert (upon observing the dialogue corpus), and the BN parameters are learned using Maximum Likelihood. This configuration is similar to the one outlined in Sect. 4.5.2.
2. *Bayesian learning method* [**BayL-BN**]: This configuration is similar to the MaxL-BN configuration, but now maximum a posteriori probability (MAP) learning is used, where the required Dirichlet priors are fixed using expert knowledge.
3. *Handcrafted method* [**HC-BN**]: In this configuration the BN parameters were hand-coded by an expert.
4. *Bayesian EM—Case 1* [**EM-C1-BN**]: In this configuration, the knowledge parameters are assumed to be missing, and the BN parameters are learned from the dialogue corpus using Bayesian EM. The priors required for this method are Dirichlet distribution coefficients which can be fixed by an expert. This makes it possible to assign varying degrees of importance to expert knowledge. As a first case, they are ignored during training in this configuration.
5. *Bayesian EM—Case 2* [**EM-C2-BN**]: This is the same configuration as EM-C1-BN, but now, the Dirichlet priors are set by an expert, giving some weight to expert knowledge during training.
6. *Bayesian EM—Case 3* [**EM-C3-BN**]: This is the same configuration as EM-C2-BN, but now, the Dirichlet priors are set and fully taken into account during training.

4.5.3.2 Experimental Results and Analysis

In order to evaluate the six simulated users, 1,000 dialogues were generated for each user, in interaction with the HIS-POMDP dialogue manager [45]. To quantify the quality of the simulations, the average number of turns as well as four dissimilarity metrics were computed (as described in Sect. 4.3): precision, recall and symmetric Kullback-Leibler dissimilarity *DS*. Note that precision and recall should be as high as possible and the Kullback–Leibler dissimilarity as low as possible. The average number of turns per simulated dialogue should be as close as possible to the average

Table 4.6 Evaluation of BN user simulations behaviours under different settings

Eval. measure	MaxL-BN	BayL-BN	HC-BN	EM-C1-BN	EM-C2-BN	EM-C3-BN
Precision	47.11	50.62	63.63	63.71	64.60	67.13
Recall	57.89	60.68	53.20	61.84	63.83	69.27
DS	0.7292	0.6712	0.8803	0.6674	0.7864	0.5288
Avg. turns	18.19	15.15	5.283	7.690	7.980	8.703

number of turns per dialogue in the corpus (which is 8.185). The evaluation results from 1,000 dialogues generated with each simulated user are presented in Table 4.6. The results indicate that the handcrafted simulator and the two simulators trained with complete data do not generate as realistic dialogues as the BN simulators trained with incomplete data using EM. Except for *DS* on the EM-C2-BN user, all scores are better for the EM simulators. In terms of recall, *DS* and number of turns, BayL-BN gives better results than MaxL-BN, which indicates that the dialogue corpus is not large enough to make accurate estimates without using priors from an expert. The HC-BN was designed to generate shorter dialogues and therefore shows a larger divergence from the corpus. These results also indicate that the training techniques with missing data are efficient, and therefore the error-prone (automatic or manual) knowledge inference is no longer required. In terms of precision, recall and dialogue length, taking the expert information into account allows us to improve the performance. In terms of *DS*, the best results are obtained with the EM-C3-BN simulator, which uses EM in combination with full expert knowledge in the form of Dirichlet priors.

4.6 Inverse Reinforcement Learning Based User Simulation

In this section, we present an approach to user simulation in which the task of simulating the behaviour of human users is perceived as a sequential decision-making problem. In deciding to generate a user act for a given dialogue context (which includes the last system act as well as the user knowledge), the user follows a long-term policy. One way to estimate the user policy from an annotated dialogue corpus is to employ reinforcement learning (RL) directly for user policy optimisation. However, RL-based optimisation requires a reward function to be defined in advance, which describes how good it is for a user simulation to perform an action in a given dialogue context. In practice, predicting the true reward function underlying the behaviour of human users turns out to be an impossible task, even for only slightly complex behaviour [1].

Therefore, the task of simulating the human users is treated as an imitation learning problem, in which IRL [25] is employed to estimate some reward function of the user. A direct RL algorithm is then used for retrieving the strategy of the user that is optimal with respect to that inferred reward function [6].

4.6.1 User Simulation as a Sequential Decision Making Problem

Similar to spoken dialogue modelling, the task of user simulation can be perceived as a sequential decision-making problem. The MDP [4] provides an efficient framework for solving sequential decision making problems. Formally, an MDP is defined as a tuple $\{S, A, P, R, \gamma\}$ where S is the state space, A is the action space, $P_{sa} : S \to \mathbb{R}$ a set of Markovian transition probabilities, $R : S \to \mathbb{R}$ the reward function and γ a discount factor weighting long-term rewards. After each transition the agent receives a reward $r_i = R(s_i)$. This section first provides an introduction to the dialogue task under study and then outlines the process of casting the task of dialogue management and user simulation as an MDP.

4.6.1.1 Description of the Dialogue Task

The dialogue problem studied in this section is a task-oriented, slot-filling dialogue system in the restaurant domain, similar to the one studied in Sects. 4.4 and 4.5, as well as in Chap. 5. In order to optimise the MDP user simulation (userMDP), an MDP dialogue manager (sdsMDP) is required [23, 42]). The dialogue state of the sdsMDP is composed of the knowledge of three slots: the *location*, *cuisine*, and *price-range* of the restaurant (one Boolean value for each slot, where "true" means the slot is filled and "false" indicates the slot is empty). There are 13 possible system acts: Ask-slot (3 actions), Explicit-confirm (3 actions), Implicit-confirm and Ask-slot value (6 actions) and finally Close-dialogue.

4.6.1.2 Casting User Simulation as an MDP

The userMDP is defined as a tuple $\{S, A, P, \gamma, R\}$. The user state is a summary of the dialogue course from a user simulation's perspective. Apart from encoding the exchange of information between the user model and the dialogue manager, the user state also includes the most recent action performed by the dialogue manager. The state representation of the userMDP consists of four variables: {System-Act}, {Slot1}, {Slot2} and {Slot3}, where the system-act takes values 0–13 corresponding to the actions defined above for the sdsMDP. Slot1, Slot2, and Slot3 take values 0–2, where 0 indicates that the slot is empty (never provided by the user), 1 indicates that the slot has been provided by the user, and 2 indicates that the slot is confirmed by the system. The action space of userMDP includes the following 10 user acts: remain silent (Silent), provide-all-values (AllSlots), provide-one-value (OneSlot: 3 actions), confirm slot value (Confirm: 3 actions), negate slot value (Negate: 3 actions) and hangup (CloseDialogue).

The solution of the userMDP is a policy π, which defines what action to perform given the user state. This policy can be estimated using RL, similar to policy optimisation for a dialogue system (which maximises the reward function specified by the system designer). However, manually specifying the non-observable reward function of the human user that is required for policy optimisation does not necessarily result in behaviour that is similar to that of human users. To overcome this problem, we explore the possibility of using IRL for user simulation.

4.6.2 Inverse Reinforcement Learning

IRL aims at learning the reward function optimised by an *expert*. This task is supposed to be accomplished using a set of observed interactions between the expert and the MDP. However, the IRL problem is ill-posed since there exist several possible reward functions that can match the expert behaviour [25]. Taking this into account, most existing IRL algorithms try to retrieve some reward function (not necessarily the true reward function) which can exhibit a behaviour similar to the expert's behaviour. An IRL-MDP is defined by the tuple $\{S,A,P,\gamma\}/R$, where $/R$ means that the reward function of the MDP is not available. Let the feature space of the reward function be defined using a vector ϕ. Then, the reward function R of the MDP can be expressed as $R_\theta(s,a) = \theta^T \phi(s,a) = \sum_{i=1}^{k} \theta_i \phi_i(s,a)$. Using this, the state-action value function, i.e. the Q-function of the MDP, can be defined as

$$Q^\pi(s,a) = E\left[\sum_{i=0}^{\infty} \gamma^i \theta^T \phi(s,a) | s_0 = s, a_0 = a\right] = \theta^T \mu^\pi(s,a), \tag{4.13}$$

where $\mu^\pi(s,a)$ is the *feature expectation* of the policy π. Feature expectation is defined as the discounted measure of features according to the state visitation frequency (based on when a state is visited in a trajectory). It provides a compact way to summarise the behaviour of the expert observed in the form of trajectories. Given a set of m trajectories (where H_i is the length of the ith trajectory) from the expert (who acts based on some unknown policy π), the feature expectation $\mu^\pi(s_0,a)$ can be estimated by:

$$\mu^\pi(s_0,a) = \frac{1}{m} \sum_{i=0}^{m-1} \sum_{t=0}^{H_i} \gamma^t \phi(s_t^i, a_t^i). \tag{4.14}$$

IRL tries to retrieve some reward function R_θ^* which can predict a behaviour (π_{predict}) that is similar to the behaviour of the expert (π_{expert}):

$$R_\theta^*(s,a) = \{J(\pi_{\text{expert}}, \pi_{\text{predict}})\},$$

where J is some dissimilarity measure between the expert behaviour and the predicted behaviour. From Eq. (4.13) it can be observed that comparing two different behaviours (policies) in terms of feature expectation is indeed comparing the

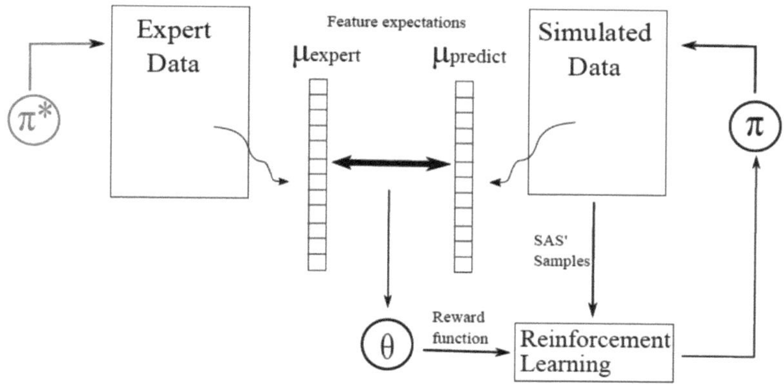

Fig. 4.7 User simulation using imitation learning

behaviours based on their Q-function(s). A greedy policy ${}_aQ(s,a)$ can be inferred for every state, and therefore, IRL-based user simulation can generalise to unseen situations which is harder to obtain from more traditional approaches which, for example, rely on assumptions about similarities of states or on smoothing of probabilities. Assuming that $||\theta|| \leq 1$ (which is not restrictive: rewards are bounded and scale invariant), one has $||Q^{\pi}|| \leq ||\mu^{\pi}||$. Thus an easy way of computing the dissimilarity between the expert behaviour π_{expert} and the predicted user behaviour π_{predict} is to compute a distance between their feature expectations, i.e., μ_{expert} and μ_{predict}:

$$J(\pi_{\text{expert}}, \pi_{\text{predict}}) = ||\mu_{\text{expert}} - \mu_{\text{predict}}||^2.$$

Imitation learning [1] focuses on learning to imitate the behaviour of the expert observed in the sample trajectories. In case of imitation learning, IRL is used as a first step to retrieve the underlying reward function of the expert, which in turn is used for policy optimisation to imitate the expert. Imitation learning can be pictorially represented as shown in Fig. 4.7.

4.6.3 User Simulation Using IRL

Let the state-action space of the userMDP be defined by a vector of features ϕ: $S \times A \rightarrow [0,1]^k$. Also it is assumed that a set of m dialogue episodes (trajectories) from human users is made available. Our primary focus is to build a user simulation which can imitate the behaviour observed in the dialogue corpus. Let us term π_{human} the policy of the human users. Feature expectation μ_{human} can be computed as shown in Eq. (4.14). In the imitation learning algorithm (Fig. 4.8), step 3 is an IRL step where, based on the dissimilarity between the human and simulated user behaviour, the reward function of the human user is estimated. It searches for the reward

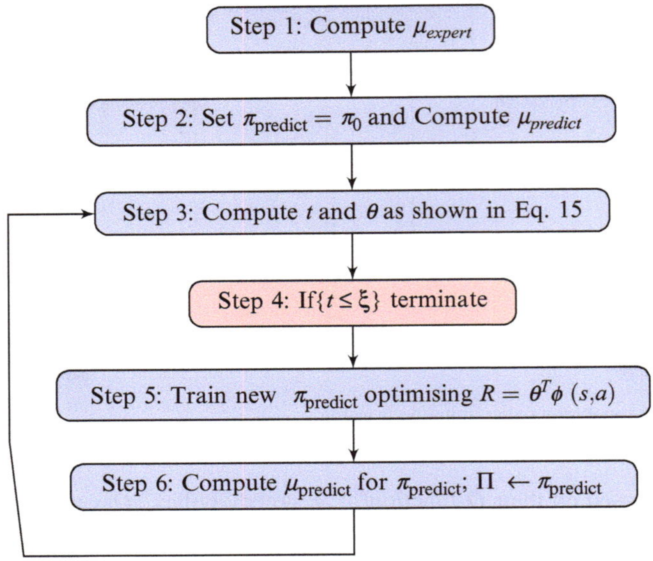

Fig. 4.8 Imitation learning via IRL algorithm

function which maximises the distance between the value of the expert and any policy computed in prior iterations [see Eq. (4.15)]. The updated reward function is used to estimate the optimal behaviour (policy for userMDP) of the simulated user in step 5. Steps 3 to 6 are performed iteratively until some convergence criteria are met, i.e., the distance between the human and simulated user behaviour is within an acceptable range ξ. Upon convergence, the algorithm outputs a set of policies Π.

$$t = \max_{\theta} \left\{ \min_{\pi_{\text{predict}\in\Pi}} \theta^T (\mu_{\text{expert}} - \mu_{\text{predict}}) \right\} \text{ s.t. } \|\theta\|^2 \leq 1. \tag{4.15}$$

In order to simulate the user one can choose the best policy from Π based on its distance with μ_{expert} or choose to use multiple policies with an associated selection probability (explained in Sect. 4.6.4). The algorithm does not guarantee to retrieve the true reward function of the expert but to retrieve some reward function which can exhibit a similar behaviour. Even though it may seem that using the state visitation frequency in the form of feature expectation is comparable to existing approaches for user simulation (such as the n-gram method), it is worth noting that the feature expectation is not directly used. It is used to predict a reward function which in turn is used to simulate the expert behaviour. Most existing ML-based approaches for user simulation focus on simulating the user at the transition level when the primary focus of this approach is to simulate user trajectories. Also IRL user simulation generalises through the value function which is non-zero everywhere.

Table 4.7 Handcrafted user behaviour

SystemAct	UserActs (probability)
Greet	Silent (0.7) AllSlots (0.3)
AskSlot	OneSlot (0.95) AllSlots (0.05)
Explicit-Conf	Confirm (1.0)
Implicit-Conf	OneSlot (0.9) Negate (0.1)
CloseDialogue	Silent (1.0)

4.6.4 Experimental Setup and Results

In order to explore the possibility of using IRL for user modelling a simple experiment is outlined here. The goal of the experiment is to learn a user model which can imitate the behaviour of a handcrafted user model (considered as the expert behaviour). For the sake of simplicity, a handcrafted dialogue policy is used for dialogue management (sdsMDP). The handcrafted dialogue strategy tries to fill and confirm all the 3 slots one after the other.

4.6.4.1 Learning to Imitate the Expert Behaviour

The user behaviour to be imitated (expert policy for userMDP) is detailed in Table 4.7. It may be recalled that the imitation learning algorithm outlined in Sect. 4.6.3 requires the feature expectation [defined in Eq. (4.14)] for a user behaviour π to be computed. One way to estimate this is to run a set of dialogue episodes between the dialogue manager (sdsMDP) and the user simulation (userMDP controlled by policy π). Using the generated data, the feature expectation of the expert user behaviour π_{expert} is computed. The computed feature expectation will be the μ_{expert} since the task here is to imitate the expert behaviour. First, a set of dialogue episodes are generated using the userMDP with a fully random action selection π_{random} and the dialogue manager. The feature expectation of the random policy can then be computed from the dialogue trajectories. Since the necessary expert behaviour is available in the form of dialogue trajectories, μ_{expert} can be computed.

The reward function of the user simulation is estimated based on the distance between the feature expectations of the expert and simulated user behaviour. At each iteration of imitation learning the intermediate reward function is used to estimate the optimal behaviour for userMDP using Least Squares Policy Iteration (LSPI) [20]. Notice that LSPI is a batch algorithm which allows learning from fixed sets of data (this is an important feature to extend the proposed method to real applications). Upon convergence (ξ set to 0.1) the best policy π^* of the userMDP (chosen from Π based on the least dissimilarity measure between π_{predict} and π_{expert}) will exhibit the expert user behaviour.

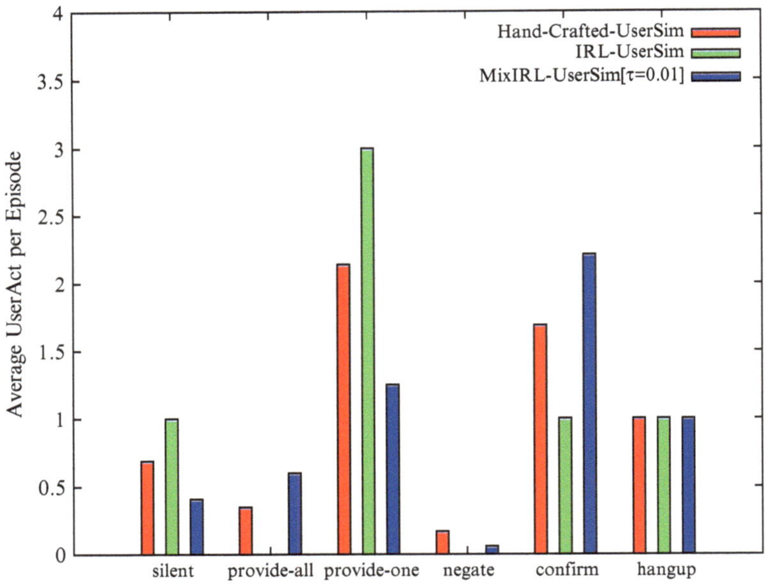

Fig. 4.9 Frequency of user actions per episode

4.6.4.2 Evaluation Results and Analysis

In order to quantify the quality of the IRL user simulation, 1,000 dialogue episodes are generated using the dialogue manager, the trained IRL user simulation and the expert user simulation (1,000 episodes each). Figure 4.9 presents the average choice of user actions taken by the two different user simulations. The behaviour of the expert user simulation is stochastic (see Table 4.7), and the behaviour of the IRL user simulation is deterministic since only one of the policies retrieved during training is used during evaluation. However, it is also possible to obtain a stochastic user behaviour by employing a *policy mixer* to pick up a different policy before starting each dialogue episode. Let the probability of choosing $\pi^i_{\text{predict}} \in \Pi$ be λ_i. The values of $\lambda_{1\ldots n}$ can be heuristically determined using a Gibbs distribution: $\lambda_i = e(-d_i/\tau)/\sum_{j=1}^{n} e(-d_j/\tau)$, where d_i is the distance between μ_{expert} and μ^i_{predict} estimated during training. During the evaluation, τ is set to 0.01 (ensures that the policies closest to π_{expert} are given more preference). Let the behaviour of the IRL user simulation which employs the policy mixer be termed as "MixIRL-UserSim".

As already suggested in Sect. 4.3.2.6, using IRL to estimate the reward function of the user provides a unique opportunity to evaluate the simulated behaviour using the reward function itself. Table 4.8 shows the average discounted reward obtained by the user simulations based on the estimated reward function. It can be observed that both the average rewards and the average dialogue lengths obtained by the IRL user simulation and the hand-crafted user simulation are very similar.

Table 4.8 Handcrafted vs. IRL user behaviour

UserSim	AvgLength	AvgReward
Hand-crafted	6.03	2.6
IRL	6.0	2.7
Mixed-IRL	5.5	2.9

A similarity measure suggested by [37] is used to compare the behaviours. This metric is used to measure the correlation between the trajectories of the expert user model, π_{hc} and the IRL user model, π_{irl}:

$$Sim(\pi_{hc}, \pi_{\mathrm{irl}}) = \frac{1}{n}\sum_{i=1}^{n} \frac{1}{1+\mathrm{rank}(\mathrm{a}_i)}$$

with n the number of user acts observed in the trajectory generated by the expert user model. Ranking of user acts (selected by π_{hc}) based on their Q-value (obtained using Q-function of the IRL used model) gives an indication of to what extent the user acts selected by the expert model correlate with the learned IRL user model. Upon comparing the expert user model with the IRL user model, the similarity measure is found to be 0.48 for IRL-UserSim and 0.46 for MixIRL-UserSim which are close to the optimal value 0.5 (for identical rankings, i.e., $\mathrm{rank}(\mathrm{a}_i)$ is 1 for all i). This confirms that the behaviour of the IRL user model correlates very well with the expert's behaviour.

4.7 Conclusion and Future Work

In this chapter, we have given an overview of user simulation techniques and evaluation methodologies. After a brief survey of metrics for evaluating user simulations, we discussed three approaches in more detail: (1) the agenda-based approach, (2) the dynamic Bayesian network approach and (3) the IRL based approach. The evaluation of the resulting user simulations showcased how different evaluation metrics can be used to quantify the effectiveness and naturalness of generated synthetic data. The primary difference between the user modelling techniques presented here and other existing user modelling techniques is the fact that the user response is generated on the basis of more information than merely the most recent system action. Considering the long-term user goal and the user's knowledge results in simulators that behave more similarly to human users.

The most recent work on agenda-based user simulation has involved its extension to be trainable on real user data whilst preserving the necessary rationality and complexity for effective training and evaluation of dialogue manager policies. The extension involved the incorporation of RDPs in the process of receiving and responding to a system act in each turn. The decisions made at these points are controlled by probability distributions defined by a set of parameters. A sample-based maximum likelihood approach to estimating these parameters from real user data in a corpus of human–machine dialogues was discussed [18]. Apart from improving the simulator by adding more RDPs and, hence, more variation in user behaviour, the evaluation should also be extended by looking at the other metrics discussed in Sect. 4.3 and, in particular, performing the ultimate test of evaluating policies trained on the trained simulator with human users (see Chap. 7). This will also show how well the simulator is suited for predicting the system performance.

Experiments on using Bayesian networks (BNs) for user simulation showed the feasibility of generating natural data when the BN parameters are estimated from a dialogue corpus. The user's knowledge of the dialogue context, which is incorporated in the Bayesian network model, is often very difficult to annotate. To tackle this problem, we have proposed to use expectation-maximisation (EM) algorithms to estimate the BN user model parameters [35]. In the future, this user model will be used to train a reinforcement learning based dialogue manager so as to optimise the dialogue strategy. Also, the extension of this user simulation technique to other tasks is envisioned. Furthermore, the possibility of online learning of the BN parameters has to be analysed, in order to continue to improve the naturalness of the synthetic data.

In a more recent approach to user simulation, the task of modelling user behaviour is cast as a MDP, and user behaviour policies are then trained from data using IRL [6]. IRL-based methods make it possible to learn complex user models with relative ease since they are based on the computation of a utility function and generating adaptive behaviours. Additionally, user models can adapt their behaviour to the environment and therefore can be seen as a step closer to human–computer interaction, where human users adapt to the system they are interacting with. Preliminary experimental results presented here showed that using IRL, it is possible to learn both deterministic and stochastic user simulation strategies. Exploring the ability of IRL based user simulation to co-adapt and to generalise (i.e., predict user responses for unseen situations) is certainly an interesting direction of future work.

Recent developments in learning dialogue management policies directly from data or online in interaction with human users suggest that simulated users might not be required in the near future [12,27]. However, the development of the dialogue manager as a whole into a framework enabling these policy optimisation approaches still relied on a lot of testing with a simulated user. Moreover, a simulated user might still be used to train an initial policy that can be deployed to real users and from then on be improved via online learning. Considering the fact that more complex dialogue problems have to be dealt with in the future, building user simulations from dialogue corpora promises to be an attractive approach. On the other hand, handcrafted user simulations may become ineffective in more complex conversational settings, even more so when considering multimodal, multi-user, situated dialogue. These, in particular, are the challenges we are currently dealing with in an ongoing human–robot interaction project [15], where the aim is to build simulated users for training policies for socially appropriate robot behaviour.

Acknowledgements The research leading to these results has received partial support from the European Cummunity's Seventh Framework Programme (FP7) under grant agreement no. 216594 (CLASSiC project), under grant agreement no. 270019 (SPACEBOOK project), and under grant agreement no. 270435 (JAMES project).

References

1. Abbeel, P., Ng, A.Y.: Apprenticeship learning via inverse reinforcement learning. In: Proceedings of the International Conference on Machine Learning (ICML), Banff, Alberta, Canada (2004)
2. Ai, H., Litman, D.: Assessing dialog system user simulation evaluation measures using human judges. In: Proceedings of the 46th meeting of the Association for Computational Linguistics, pp. 622–629. Columbus, OH (2008)
3. Anderson, T.: On the distribution of the two-sample Cramér-von Mises criterion. Annals of Mathematical Statistics **33**(3), 1148–1159 (1962)
4. Bellman, R.: A markovian decision process. Journal of Mathematics and Mechanics **6**, 679–684 (1957)
5. Bos, J., Klein, E., Lemon, O., Oka, T.: DIPPER: Description and Formalisation of an Information-State Update Dialogue System Architecture. In: 4th SIGdial Workshop on Discourse and Dialogue, pp. 115–124. Sapporo (2003)
6. Chandramohan, S., Geist, M., Lefèvre, F., Pietquin, O.: User Simulation in Dialogue Systems using Inverse Reinforcement Learning. In: Proceedings Interspeech 2011, Florence, Italy, August 2011
7. Cheyer, A., Martin, D.L.: The open agent architecture. Autonomous Agents and Multi-Agent Systems **40**(1/2), 143–148 (2001)
8. Cramer, H.: On the composition of elementary errors. second paper: Statistical applications. Skandinavisk Aktuarietidskrift **11**, 171–180 (1928)
9. Cuayáhuitl, H., Renals, S., Lemon, O., Shimodaira, H.: Human-computer dialogue simulation using hidden markov models. In: Proceedings of the Automatic Speech Recognition Workshop (ASRU), Cancun, Mexico (2005)
10. Doddington, G.: Automatic evaluation of machine translation quality using n-gram co-occurrence statistics. In: Proceedings of the Human Language Technology Conference (HLT), San Diego, CA (2002)
11. Eckert, W., Levin, E., Pieraccini, R.: User modeling for spoken dialogue system evaluation. In: Proceedings of the Automatic Speech Recognition Workshop (ASRU), Santa Barbara, CA, December 1997
12. Gasic, M., Jurcicek, F., Thomson, B., Yu, K., Young, S.: On-line policy optimisation of spoken dialogue systems via live interaction with human subjects. In: Proceedings of the Automatic Speech Recognition Workshop (ASRU), Waikoloa, HI (2011)
13. Georgila, K., Henderson, J., Lemon, O.: User simulation for spoken dialogue systems: Learning and evaluation. In: Proceedings International Conference on Spoken Language Processing (Interspeech/ICSLP), Pittsburgh, PA (2006)
14. Götze, J., Scheffler, T., Roller, R., Reithinger, N.: User simulation for the evaluation of bus information systems. In: Proceedings IEEE Spoken Language Technology Workshop (SLT), Berkeley, CA, December 2010
15. Foster, M.E., Keizer, S., Wang, Z., Lemon, O.: Machine learning of social states and skills for multi-party human-robot interaction. In: Proceedings ECAI Workshop on Machine Learning for Interactive Systems: Bridging the Gap Between Language, Motor Control and Vision, Montpellier (MLIS), France, (2012)

16. Jung, S., Lee, C., Kim, K., Jeong, M., Geunbae Lee, G.: Data-driven user simulation for automated evaluation of spoken dialogue systems. Computer Speech and Language **23**, 479–509 (2009)
17. Keizer, S., Gašić, M., Mairesse, F., Thomson, B., Yu, K., Young, S.: Modelling user behaviour in the HIS-POMDP dialogue manager. In: Proceedings of SLT, 2008
18. Keizer, S., Gašić, M., Jurčíček, F., Mairesse, F., Thomson, B., Yu, K., Young, S.: Parameter estimation for agenda-based user simulation. In: Proceedings of the Annual SIGdial Meeting on Discourse and Dialogue, Tokyo, Japan, September 2010
19. Kullback, S., Leibler, R.: On information and sufficiency. Annals of Mathematical Statistics **22**, 79–86 (1951)
20. Lagoudakis, M.G., Parr, R.: Least-squares policy iteration. Journal of Machine Learning Research **4**, 1107–1149 (2003)
21. Lee, A., Przybocki, M.: NIST 2005 machine translation evaluation official results. official release of automatic evaluation scores for all submissions, August 2005
22. Lemon, O., Liu, X., Shapiro, D., Tollander, C.: Hierarchical Reinforcement Learning of Dialogue Policies in a Development Environment for Dialogue Systems: REALL-DUDE. In: Proceedings of the 10th SemDial Workshop on the Semantics and Pragmatics of Dialogue (BRANDIAL), Potsdam, Germany (2006)
23. Levin, E., Pieraccini, R.: Using Markov Decision Process for learning dialogue strategies. In: IEEE International Conference on Acoustics, Speech and Signal Processing (ICASSP), Seattle, WA (1998)
24. Levin, E., Pieraccini, R., Eckert, W.: A stochastic model of human-machine interaction for learning dialogue strategies. IEEE Transactions on Speech and Audio Processing **8**(1), 2000
25. Ng, A.Y., Russell, S.: Algorithms for inverse reinforcement learning. In: Proceedings of 17th International Conference on Machine Learning (ICML), Stanford, CA (2000)
26. Papineni, K., Roukos, S., Ward, T., Zhu, W.J.: BLEU: A method for automatic evaluation of machine translation. In: Proceedings of the 40th Annual Meeting on Association for Computational Linguistics (ACL), Philadelphia, PA (2002)
27. Pietquin, O., Geist, M., Chandramohan, S., Frezza-Buet, H.: Sample-Efficient Batch Reinforcement Learning for Dialogue Management Optimization. ACM Transactions on Speech and Language Processing **7**(3), 1–21 (2011)
28. Pietquin, O.: A probabilistic description of man-machine spoken communication. In: Proceedings of the IEEE International Conference on Multimedia & Expo (ICME), Amsterdam, The Netherlands (2005)
29. Pietquin, O.: Consistent goal-directed user model for realistic man-machine task-oriented spoken dialogue simulation. In: Proceedings of the IEEE International Conference on Multimedia and Expo (ICME), Toronto, Canada, July 2006
30. Pietquin, O., Dutoit, T.: A probabilistic framework for dialogue simulation and optimal strategy learning. IEEE Transactions on Audio, Speech and Language Processing **14**(2), 589–599 (2006)
31. Pietquin, O., Hastie, H.: A survey on metrics for the evaluation of user simulations. The Knowledge Engineering Review, 2013
32. Rieser, V.: Bootstrapping Reinforcement Learning-based Dialogue Strategies from Wizard-of-Oz data. PhD thesis, Saarland University, Department of Computational Linguistics, July 2008
33. Rieser, V., Lemon, O.: Simulations for learning dialogue strategies. In: Proceedings of Interspeech 2006, Pittsburg, PA (2006)
34. van Rijsbergen, C.J.: Information Retrieval, 2nd edn. Butterworths, London, UK (1979)
35. Rossignol, S., Ianotto, M., Pietquin, O.: Training a BN-based user model for dialogue simulation with missing data. In: Proceedings of the International Joint Conference on Natural Language Processing (IJCNLP), Chiang Mai, Thailand (2011)
36. Rossignol, S., Pietquin, O., Ianotto, M.: Grounding Simulation in Spoken Dialog Systems with Bayesian Networks. In: Proceedings of the International Workshop on Spoken Dialogue Systems (IWSDS 2010), Gotemba, Japan, October 2010

37. Schatzmann, J., Stuttle, M.N., Weilhammer, K., Young, S.: Effects of the user model on simulation-based learning of dialogue strategies. In: Proceedings of ASRU'05, 2005
38. Schatzmann, J., Weilhammer, K., Stuttle, M., Young, S.: A survey of statistical user simulation techniques for reinforcement learning of dialogue management strategies. The Knowledge Engineering Review **21**(2), 97–126 (2006)
39. Schatzmann, J., Thomson, B., Young, S.: Statistical user simulation with a hidden agenda. In: Proceedings of the Annual SIGDIAL Meeting on Discourse and Dialogue, pp. 273–282. Antwerp, Belgium (2007)
40. Schatzmann, J., Thomson, B., Weilhammer, K., Ye, H., Young, S.: Agenda-based user simulation for bootstrapping a POMDP dialogue system. In: Proceedings of the Annual Conference of the North American Chapter of the Association for Computational Linguistics (NAACL-HLT), Rochester, NY (2007)
41. Scheffler, K., Young, S.: Corpus-based dialogue simulation for automatic strategy learning and evaluation. In: Proceedings of the NAACL Workshop on Adaptation in Dialogue, Pittsburgh, PA (2001)
42. Singh, S., Kearns, M., Litman, D., Walker, M.: Reinforcement learning for spoken dialogue systems. In: Solla, S., Leen, T., Müller, K. (eds.) Advances in Neural Information Processing Systems (NIPS), MIT Press (2000)
43. Syed, U., Williams, J.D.: Using automatically transcribed dialogs to learn user models in a spoken dialog system. In: Proceedings of the Annual Meeting of the Association for Computational Linguistics (ACL) with the Human Language Technology Conference (HLT), Columbus, OH (2008)
44. Williams, J.: Evaluating user simulations with the Cramér-von Mises divergence. Speech Communication **50**, 829–846 (2008)
45. Young, S., Gašić, M., Keizer, S., Mairesse, F., Thomson, B., Yu, K.: The Hidden Information State model: a practical framework for POMDP based spoken dialogue management. Computer Speech and Language **24**(2), 150–174 (2010)
46. Zukerman, I., Albrecht, D.: Predictive statistical models for user modeling. User Modeling and User-Adapted Interaction **11**, 5–18 (2001)

Chapter 5
Optimisation for POMDP-Based Spoken Dialogue Systems

Milica Gašić, Filip Jurčíček, Blaise Thomson, and Steve Young

5.1 Introduction

Spoken dialogue systems (SDS) allow users to interact with a wide variety of information systems using speech as the primary, and often the only, communication medium. The principal elements of an SDS are a speech understanding component which converts each spoken input into an abstract semantic representation called a user dialogue act (see Chap. 3), a dialogue manager which responds to the user's input and generates a system act a_t in response, and a message generator which converts each system act back into speech (see Chap. 6). At each turn t, the system updates its state s_t, and based on a policy π, it determines the next system act $a_t = \pi(s_t)$. The state consists of the variables needed to track the progress of the dialogue and the attribute values (often called slots) that determine the user's requirements. In conventional systems, as discussed in Chap. 8, the policy is usually defined by a flow chart with nodes representing states and actions and arcs representing user inputs.

Despite steady progress over the last few decades in speech recognition technology, the process of converting conversational speech into words still incurs word error rates in the range 15% to 30% in realistic operating environments. Systems which interpret and respond to spoken commands must therefore implement dialogue strategies which account for the unreliability of the input and provide error checking and recovery mechanisms. As a consequence, conventional deterministic flowchart-based systems are typically expensive to build and fragile in operation.

During the last few years, a new approach to dialogue management has emerged based on the mathematical framework of partially observable Markov decision processes (POMDPs) [21, 32, 33]. This approach assumes that dialogue evolves

M. Gašić • F. Jurčíček • B. Thomson • S. Young (✉)
Engineering Department, Cambridge University, Trumpington Street,
Cambridge, CB2 1PZ , UK
e-mail: mg436@eng.cam.ac.uk; sjy11@cam.ac.uk

O. Lemon and O. Pietquin (eds.), *Data-Driven Methods for Adaptive Spoken Dialogue Systems*, DOI 10.1007/978-1-4614-4803-7_5,

as a Markov process i.e. starting in some initial state s_0, each subsequent state is modelled by a transition probability: $p(s_t|s_{t-1},a_{t-1})$. The state s_t is not directly observable reflecting the uncertainty in the interpretation of user utterances; instead, at each turn, the system can observe only the noisy interpretation of the user input o_t with probability $p(o_t|s_t)$. The transition and observation probability functions are represented by a suitable stochastic model, called here the *dialogue model* $\mathcal{M}$.

The decision as to which action to take at each turn is determined by a second stochastic model, called here the *policy model* $\mathcal{P}$. As the dialogue progresses, a reward is assigned at each step designed to reflect the desired characteristics of the dialogue system. The overall goal is to maximise the expected accumulated sum of these rewards by optimising the dialogue model $\mathcal{M}$ and policy model $\mathcal{P}$, either on-line through interaction with users or off-line from a corpus of dialogues collected within the same domain. The representation of these models can be either parametric or non-parametric.

The ultimate performance of a statistical spoken dialogue therefore depends on how effectively the two models $\mathcal{M}$ and $\mathcal{P}$ can be optimised and the purpose of this chapter is to present recent advances in this area developed within the Dialogue Systems Group at Cambridge. For background, Sect. 5.2 outlines in a little more detail the basic framework of POMDP-based dialogue systems and Sect. 5.3 explains how a tractable stochastic dialogue model can be parameterised and implemented using dynamic Bayesian networks [27]. The remaining sections focus on optimisation techniques. Section 5.4 explains how a more general form of inference called expectation propagation (EP), which can be viewed as a form of expectation-maximisation, can be used for both belief tracking and parameter optimisation [17, 28]. Section 5.5 explains how natural actor-critic reinforcement learning can be used to optimise the policy parameters $\mathcal{P}$ [19, 26] and how, with a simple extension, it can also be used to optimise the dialogue model parameters $\mathcal{M}$ [14]. Finally, Sect. 5.6 addresses the problem of fast online policy optimisation using Gaussian processes as a non-parametric policy model [8, 10, 20]. Section 5.7 wraps up with some general conclusions and pointers to future work.

5.2 POMDP-Based Dialogue Management

The first key idea of POMDP-based dialogue management is that instead of computing the most likely state at each turn, the system tracks the probability of all states. The posterior distribution over states is called the belief state b_t, and it is updated each turn using the current input observation o_t from the user. This update is referred to as *belief monitoring* or *belief tracking* and it is computed as follows:

$$b_t = p(s_t|h_t;\mathcal{M}) = kp(o_t|s_t;\mathcal{M}) \sum_{s_{t-1}} p(s_t|s_{t-1},a_{t-1};\mathcal{M})b_{t-1}, \tag{5.1}$$

where $h_t = \{a_0,o_1,a_1,o_2,\ldots,a_{t-1},o_t\}$ is the dialogue history, $\mathcal{M}$ is the chosen dialogue model and k is a normalising constant. Whilst simple in theory, the practical implementation of the Markov model underlying this belief monitoring process is complex. The belief state b_t is continuous and has very high dimension which makes direct implementation impossible. One solution is to group equivalent states into partitions and maintain only the N-best most likely partitions as is done in the HIS system [34]. While such a representation enables a more detailed expression of the dialogue state [11], it requires a non-parametric modelling of the transition and observation probabilities [12]. Another approach is to factorise the model into smaller components and assume the majority of components are independent. This allows the model then to be parameterised, $\mathcal{M} = \mathcal{M}(\tau)$. This is typically done with multinomial distributions, which then naturally leads to an implementation based on dynamic Bayesian networks [27]. This approach is explained further in the next section since it underpins expectation propagation methods of parameter optimisation. It should be stressed, however, that the policy optimisation techniques in Sects. 5.5 and 5.6 are applicable to virtually all statistical dialogue systems since they make few assumptions about the underlying dialogue models.

The second key idea of POMDP-based dialogue management is that rather than mapping states into actions as in a conventional system, the policy of a POMDP maps whole belief states into actions. Thus, the decision as to what to do next is not simply dependent on a single assumed state; instead, it takes account of the full distribution across all states. This mapping can be modelled deterministically as in $a_t = \pi(b_t)$, but since b_t is a high-dimension continuous vector, the same tractability issues encountered with the implementation of the dialogue model apply. An alternative approach is to represent the mapping by a stochastic model $\mathcal{P}$

$$a_t \sim \pi(a|b_t;\mathcal{P}). \tag{5.2}$$

This stochastic model can then be parameterised $\mathcal{P} = \mathcal{P}(\theta)$, which allows a wide range of function approximation techniques to be applied. For example, the features of belief space which directly impact each possible system act a_i can be encoded in a basis function $\phi_{a_i}(b_t)$. A weighted set of these basis functions can then be cast into a probability via a softmax function as in

$$\pi(a_t|b_t;\theta) = \frac{e^{\theta\cdot\phi_{a_t}(b_t)}}{\sum_a e^{\theta\cdot\phi_a(b_t)}}, \tag{5.3}$$

where $\cdot$ denotes a scalar product. Section 5.5 explains how the parameters θ of this model can be efficiently optimised using a natural actor-critic framework.

However, whereas for slot-based dialogue systems it is often easy to make assumptions about the shape of the dialogue model distributions, designing an accurate parameterised model for a policy may be hard. An alternative approach therefore is to use a non-parametric model for the policy. Section 5.6 explains how Gaussian processes can be used for this purpose.

Unlike conventional classification tasks, the objective of a dialogue system is to achieve some long-term goal through a sequence of planned actions. This is an

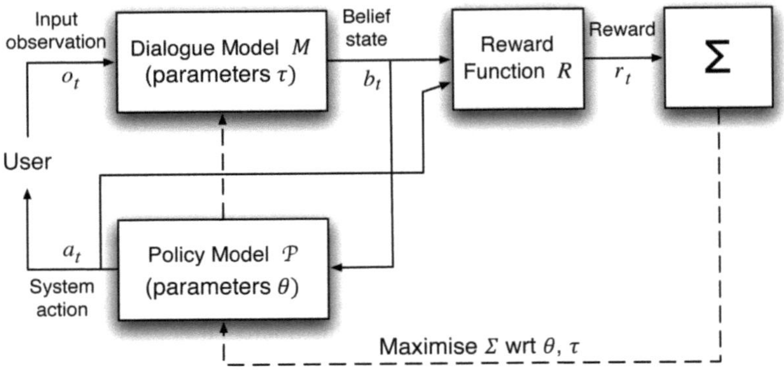

Fig. 5.1 General framework for parameter optimisation in a POMDP-based spoken dialogue system. The input/output interaction may be with a real user or a simulated user

example of planning under uncertainty and optimisation can be performed using reinforcement learning. Each dialogue turn is assigned a reward r_t based on the current state and action a_t:

$$r_t = r(b_t, a_t) = \sum_{s_t} b_t(s_t) r(s_t, a_t). \tag{5.4}$$

The goal of reinforcement learning is then to maximise the expected discounted value of the total accumulated reward:

$$R = \mathcal{E}\left\{\sum_{t=1}^{T} \gamma^{t-1} r_t\right\}, \tag{5.5}$$

where T is the length of the dialogue and $\gamma \leq 1$ is a discount factor included to allow the present value of future rewards to be discounted. In practical limited domain SDS such as information inquiry, the reward function will typically have the form

$$r_t = \begin{cases} -1 & \text{if } t < T, \\ +20 & \text{if } t = T \text{ and dialogue successful,} \\ 0 & \text{if } t = T \text{ and dialogue unsuccessful.} \end{cases} \tag{5.6}$$

In this case, it would be normal to set $\gamma = 1$.

The above sets out the basic framework of a POMDP-based spoken dialogue system and the main elements are summarised in Fig. 5.1. A dialogue model $\mathcal{M}$ maintains a distribution b_t over all possible dialogue states, updating it each turn in response to each new observation o_t. A dialogue policy $\mathcal{P}$ then determines the best system action a_t to take given the current belief state b_t via a mapping $a_t = \pi(b_t)$. At each turn, a reward r_t is generated and accumulated. The overall objective is to maximise the expected value of this accumulated reward.

Before addressing the issue of how to optimise the dialogue model $\mathcal{M}$ and the policy model $\mathcal{P}$, it will be helpful to give a little more detail on how these models are typically implemented.

5.3 A Dynamic Bayesian Network Dialogue Model

One strategy that can be used to simplify the dialogue model is to factorise the state into a collection of sub-components. For example, many POMDP dialogue models factorise the state at a particular point in time, s_t, into the user's goal, g_t, the true user action, u_t, and a dialogue history, h_t, as suggested by [31]. In some cases, these sub-components can be further factorised according to a collection of *concepts* $c \in \mathcal{C}$, so that the goal is made up of sub-goals $g_{c,t}$, and similarly $h_t = \{h_{c,t}\}_{c \in \mathcal{C}}$ (see Fig. 5.2).

The CAMINFO restaurant system, which will be used for experiments later in the chapter, provides a good example of how this approach can be used in practice. The task of this system is to provide information about restaurants in Cambridge with users speaking to the system over the phone. Figure 5.2 enumerates the different concepts in the system.

As mentioned in Sect. 5.2, the naïve approach of simply implementing Eq. 5.1 to update the system's beliefs soon becomes intractable. The factorisation described above, along with some conditional independence assumptions, enables the use of standard machine learning algorithms to update the beliefs. This is done as follows.

First, the variables of interest are represented in a Bayesian network [2]. Each node of the network denotes a random variable of interest, and edges in the network encode conditional independence assumptions. The assumption made is that the joint distribution of all nodes factorises into the product of the distribution of each node given its parents in the graph. A Bayesian network representation of part of the CAMINFO system is given in Fig. 5.3. In the CAMINFO system, the sub-histories

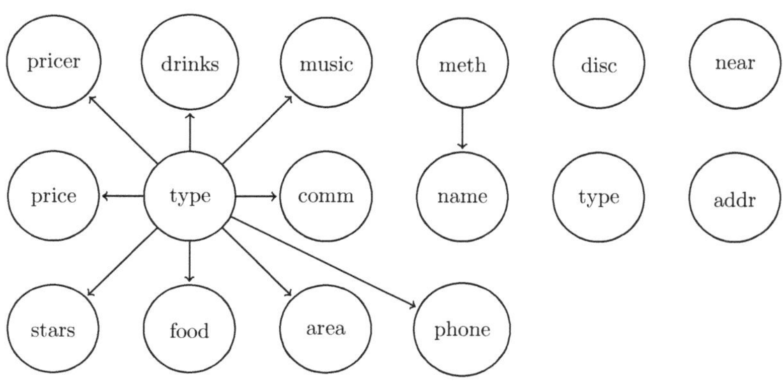

Fig. 5.2 Concepts in the CAMINFO restaurant system

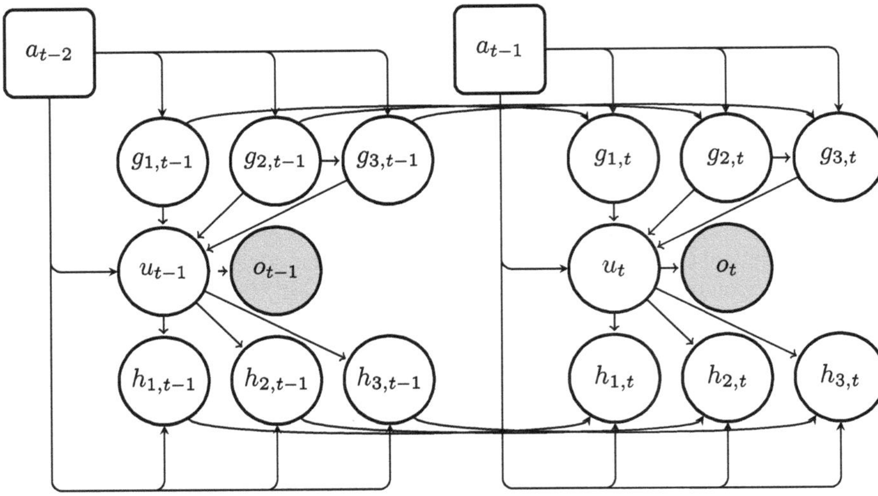

Fig. 5.3 Bayesian network for three concepts in the CAMINFO system

are assumed to be dependent on only their previous value, the true user action and the system's action. The observation is dependent on only the user act. The user act depends on all the sub-goals and the last system action while the sub-goals depend on their previous value, the system action and optionally other goals in the network. The dependencies between the goals are shown by the arrows in Fig. 5.2.

The probability functions are grouped into two types. The first type is called the *mostly constant factor* type, and it is used for the probabilities of goal nodes and history nodes. Suppose a node is labelled X_t and a collection of "special" values are defined, $x_1, x_2, \ldots, x_k$, along with a collection of "parent classes". Example "special values" are `dontcare` and `N/A` which are used to represent goals where the user does not care about the value and where the concept is not applicable. In the CAMINFO system, the parent class allows one to define different probability structures for different situations (based on the system action and the node's other parents), but to still allow for significant amounts of parameter tying. An example of such parent classes arises in the case when the system has informed that no venue has a particular concept value. Let the parents of X_t, excluding the node's previous value, be denoted by $par'(X_t)$ and let $\rho(par'(X_t))$ denote the associated parent class. The probabilities are then defined in such a way that the "special values" have distinct probabilities but all remaining values are considered equivalent in order to tie the parameters. Formally, the mostly constant probability functions are defined as follows, with $i, j \in \{1, 2, \ldots, k\}$ and $x, y \notin \{x_1, x_2, \ldots, x_k\}$ and $x \neq y$:

$$p(X_t = x_j | X_{t-1} = x_i, \rho(par'(X_t)) = \rho') = \boldsymbol{\tau}_{\rho',i,j}$$
$$p(X_t = x | X_{t-1} = x_i, \rho(par'(X_t)) = \rho') = \boldsymbol{\tau}_{\rho',i,k+1}$$

$$
\begin{aligned}
p(X_t = x_j | X_{t-1} = x, \rho(par'(X_t)) = \rho') &= \boldsymbol{\tau}_{\rho',k+1,j} \\
p(X_t = x | X_{t-1} = x, \rho(par'(X_t)) = \rho') &= \boldsymbol{\tau}_{\rho',k+1,k+1} \\
p(X_t = y | X_{t-1} = x, \rho(par'(X_t)) = \rho') &= \boldsymbol{\tau}_{\rho',k+1,k+2}.
\end{aligned}
$$

This structure of probability factors allows parameters that are expected to be similar to be tied together. For example, in modelling the changes in the user's goal for food, the different types of food will not be included in the list of "special values". As a result, the probability of the user's food goal changing in one turn to the next from, say, Indian to Italian is always equal to the probability of changing between any other pair of food types such as Chinese to French. Similarly, the probability of the user's goal staying Indian will be no different to the probability of the goal staying the same for any other food type.

The second form of probability factor is used to define the probability of the user action given the parent goals and system action. Only two probabilities are differentiated, depending on whether the goals are acceptable pre-conditions for the user action given the system's prompt [15]. For example, if the user has said "I want Chinese food" then goals which include `food=Chinese` are acceptable but goals which have `food=`x for other values of x would not be acceptable. Formally, the probability is defined as

$$
p(u|g,a) = \begin{cases} \tau_1 & \text{If } g \text{ is acceptable as a set of pre-conditions} \\ \tau_2 & \text{otherwise.} \end{cases} \tag{5.7}
$$

With the probability network now defined, one can use standard algorithms such as loopy belief propagation to update the beliefs in these variables after every turn [2]. An extension of loopy belief propagation, called expectation propagation, can even provide estimates of the parameters of the model from data [17], and the use of this algorithm for updating and parameter learning will be explained in the next section.

5.4 Expectation Propagation

Expectation propagation (EP) works by approximating the joint distribution as a simpler factorised distribution. Any marginals that are required can then be easily computed by summing out the appropriate variables. The starting point of EP is therefore the joint distribution.

The joint distribution of all variables, X, in the network can be written as a product of probability factors, $p(X) = \prod_f p_f(X)$, with f indexing the factors. Each factor gives the probability of a variable given its parents in the Bayesian network. The collection of variables linked to factor f is denoted X_f. The joint can therefore be written $p(X) = \prod_f p_f(X_f)$. When a collection of variables is observed, the joint posterior distribution is again proportional to $\prod_f p_f(X_f)$, with observed variables replaced by their observed value. Expectation propagation attempts to find an approximation to this posterior, $q(X) = \prod q_f(X_f) \approx p(X)$. In this case, a factorised approximation is used, where each factor is further factorised: $q_f(X_f) = \prod_j q_f(x_j)$, with j indexing all variables and $q_f(x_j)$ constant for variables not appearing in the factor.

EP solves for this approximation one factor at a time. A particular factor, $\tilde{f}$, is chosen and all other factors are fixed. One must then find $q_{\tilde{f}}(X_{\tilde{f}}) = \prod_j q_{\tilde{f}}(x_j)$ to minimise the Kullback-Leibler (KL) divergence $KL(p||q^{\backslash\tilde{f}} q_f)$, where

$$q^{\backslash\tilde{f}}(X_{\tilde{f}}) \propto \prod_{f \neq \tilde{f}} q_f(X_f).$$

The function $q^{\backslash\tilde{f}}(X_{\tilde{f}})$ denotes the *cavity distribution*, obtained by multiplying all approximations except for $\tilde{f}$. The cavity distribution as a function of a single variable x_j is similarly defined as

$$q^{\backslash\tilde{f}}(x_j) \propto \prod_{f \neq \tilde{f}} q_f(x_j). \tag{5.8}$$

The function $q^{\backslash\tilde{f}} q_{\tilde{f}}$ is called the *target function*.

In the case here, all the probability factors to be approximated describe the probability of a discrete output variable given a collection of discrete input variables, and a collection of parameter vectors. For clarity, the effect of the parent category, ρ, and the index of the slot, i, will be omitted, and the simplest case of a goal dependent on only its previous value is presented. Note that this exposition uses the general form of probability function instead of the mostly constant probability functions presented in the previous section.

The chosen probability factor, $\tilde{f}$, has the form

$$p(g_t = j | g_{t-1} = k) = \tau_{j,k}$$

and hence approximating functions $q_{\tilde{f}}(g_t)$, $q_{\tilde{f}}(g_{t-1})$, and $q_{\tilde{f}}(\boldsymbol{\tau}_j)$ must be found. All $q_f(\boldsymbol{\tau}_j)$ approximations are constrained to the Dirichlet distribution, with the parameters denoted by $\boldsymbol{\alpha}_{f,j}$. The approximations for the other factors are fixed and the cavity distributions for the variables are defined as per (5.8). In the case of the discrete variables g_t and g_{t-1}, the cavity distributions are computed by multiplying all factor approximations except for $\tilde{f}$. The cavity distribution for the parameters $\boldsymbol{\tau}_j$ is a product of continuous distributions. For N_f different factors, the cavity distribution is the Dirichlet distribution with parameters

$$\boldsymbol{\alpha}_j^{\backslash\tilde{f}} = \sum_{f \neq \tilde{f}} \boldsymbol{\alpha}_{f,j} - (N_f - 1)\mathbf{1}. \tag{5.9}$$

Note that when the parameters $\boldsymbol{\tau}_j$ do not appear in a factor, the approximation is constant and the vector of approximation parameters, $\boldsymbol{\alpha}_{f,j}$, equals the vector of ones, $\mathbf{1}$.

Given the cavity distributions, one can show that the discrete approximating functions that minimize $\mathrm{KL}(p||q^{\backslash\tilde{f}}q_{\tilde{f}})$ are [25]

$$q_{\tilde{f}}(g_t) \propto \sum_{g_{t-1}} q^{\backslash\tilde{f}}(g_{t-1})\mathbb{E}(\boldsymbol{\tau}_{g_{t-1},g_t}|q^{\backslash\tilde{f}}(\boldsymbol{\tau}_{g_{t-1}})), \tag{5.10}$$

$$q_{\tilde{f}}(g_{t-1}) \propto \sum_{g_t} q^{\backslash\tilde{f}}(g_t)\mathbb{E}(\boldsymbol{\tau}_{g_{t-1},g_t}|q^{\backslash\tilde{f}}(\boldsymbol{\tau}_{g_{t-1}})), \tag{5.11}$$

where the expectations are taken over $q^{\backslash\tilde{f}}$.

It can be shown that to minimise the KL divergence, the set of parameters for the target function, denoted $\boldsymbol{\alpha}_j^*$, must satisfy the following equation for every k [25]:

$$\Psi(\alpha_{jk}^*) - \Psi\left(\sum_{l=1}^{N_\alpha} \alpha_{jl}^*\right) = c_{jk}, \tag{5.12}$$

where N_α denotes the number of values,

$$c_{jk} = \Psi(\alpha_{jk}^{\backslash\tilde{f}}) - \Psi\left(\sum_{l=1}^{N_\alpha} \alpha_{jl}^{\backslash\tilde{f}}\right) + \frac{w_{jk}}{\alpha_{jk}^{\backslash\tilde{f}}} - \frac{1 - w_{j0}}{\sum_{l=1}^{N_\alpha} \alpha_{jl}^{\backslash\tilde{f}}}, \tag{5.13}$$

$\Psi(z)$ is the digamma function,

$$\Psi(z) = \frac{\mathrm{d}}{\mathrm{d}z} \log \Gamma(z), \tag{5.14}$$

and the w_{jk} are weights ($\sum_k w_{jk} = 1$)

$$w_{j0} \propto \sum_{j' \neq j} q^{\backslash\tilde{f}}(g_{t-1} = j'), \tag{5.15}$$

$$w_{jk} \propto q^{\backslash\tilde{f}}(g_{t-1} = j)q^{\backslash\tilde{f}}(g_t = k)\frac{\alpha_{jk}^{\backslash\tilde{f}}}{\sum_{l=1}^{N_\alpha} \alpha_{jl}^{\backslash\tilde{f}}}. \tag{5.16}$$

Various methods are possible for solving Eq. 5.12. The approach used here is taken from Sect. 3.3.3. of [18]. Let $\Delta = \Psi(\sum_{k=1}^{N_\alpha} \alpha_{ik}^*)$ and make α_{ij}^* the subject of the formula in Eq. 5.12:

$$\alpha_{jk}^* = \Psi^{-1}(c_{jk} + \Delta). \tag{5.17}$$

Summing over k and taking both sides as arguments for the Ψ function gives,

$$\Delta = \Psi\left(\sum_{l=1}^{N_\alpha} \alpha_{jl}^{*}\right) = \Psi\left(\sum_{l=1}^{N_\alpha} \Psi^{-1}(c_{jl} + \Delta)\right). \tag{5.18}$$

One can now solve for Δ using Newton's method and use (5.17) to obtain the $\boldsymbol{\alpha}_j^*$ parameters. The desired approximating function parameters are then calculated as

$$\boldsymbol{\alpha}_{\tilde{f},j} = \boldsymbol{\alpha}_j^* - \boldsymbol{\alpha}_j^{\setminus \tilde{f}}. \tag{5.19}$$

The special forms of probability factor described in Sect. 5.3 use essentially the same update, though the computation can be simplified because of their special structure. Details can be found in [25]. The full algorithm operates by repeatedly choosing a factor to update, computing the cavity distributions in terms of the current approximations (5.8) and (5.9) and then updating the current approximating functions as per (5.10), (5.11) and (5.19). Similar to belief propagation, the process is repeated until changes in the approximating functions fall below a threshold.

5.5 Policy Gradient Methods

In Sect. 5.2, it was noted that some form of approximation is necessary to obtain a tractable policy model, and in the example provided, a stochastic policy was approximated by a softmax function (5.3) where the policy parameters θ are weights in the linear combination of a set of basis functions ϕ.

One of the most successful approaches to estimating the parameters of such a model depends on the use of policy gradient methods which repeatedly estimate a gradient of the expected reward (5.5) and take a small step in the gradient direction in order to increase the expected reward. The computation of the gradient is built around a Monte Carlo algorithm which estimates the gradient from a finite number of dialogues generated in interaction with a real user or a user simulator (see [23] and Chap. 7.2.3). Under reasonable assumptions, such an approach will converge to a local maximum of the expected reward. The core of a policy gradient algorithm relies on the availability of analytic solutions to computing the gradient of the reward with respect to the policy parameters.

Policy gradient methods are best explained in terms of the complete dialogue history H_t which consists of the observed dialogue history h_t together with the unobserved dialogue states: $H_t = \{s_0, a_0, o_1, s_1, \ldots, a_{t-1}, o_t, s_t\}$. Given this definition, the objective function in (5.5) can be expressed as the expected reward over all trajectories:

$$R(\theta) = \int p(H;\theta) R(H) \mathrm{d}H, \tag{5.20}$$

where $R(H)$ is the expected reward accumulated along the complete history H and $p(H;\theta)$ is the probability of the complete history H given the policy parameters θ.

Using "the log likelihood-ratio trick" [30] and Monte Carlo approximation using N sampled dialogues, the gradient can be estimated as follows

$$\nabla R(\theta) \approx \frac{1}{N} \sum_{n=1}^{N} \nabla \log p(H_n;\theta) R(H_n). \tag{5.21}$$

By definition, the probability $p(H;\theta)$ is the product of the probabilities of all actions, observations and state transitions along the complete history H; therefore, the probability of the complete history is as follows:

$$p(H;\theta) = p(s_0) \prod_{t=1}^{T} p(o_t|s_t) p(s_t|a_{t-1}, s_{t-1}) \pi(a_{t-1}|b_t;\theta), \tag{5.22}$$

where $p(s_0)$ is the initial state distribution. Consequently, the log-gradient can be written in the form

$$\nabla \log p(H;\theta) = \sum_{t=0}^{T-1} \nabla \log \pi(a_t|b_t;\theta) \tag{5.23}$$

since the state observation and transition probabilities do not depend on θ. Substituting this into (5.21) gives

$$\nabla R(\theta) \approx \frac{1}{N} \sum_{n=1}^{N} \sum_{t=0}^{T_n-1} \nabla \log \pi(a_t^n|b_t^n;\theta) R(H_n). \tag{5.24}$$

Note that the gradient now depends only on observed variables. To obtain a closed form solution of (5.24), the policy π must be differentiable with respect to θ. Conveniently, the logarithm of the softmax function (5.3) is differentiable with respect to θ.

To lower the variance of the estimate of the gradient, a constant baseline, B, can be introduced into (5.24) without introducing any bias [30]

$$\nabla R(\theta) \approx \frac{1}{N} \sum_{n=1}^{N} \sum_{t=0}^{T_n-1} \nabla \log \pi(a_t^n|b_t^n;\theta)^T (R(H_n) - B). \tag{5.25}$$

The gradient defined in (5.25) still cannot be used directly since it includes the expected reward $R(H)$ which is not directly observable. However, it is possible to approximate $R(H)$ by a linear function parameterised by a vector w as follows

$$R(H) \approx R(H;w) = \sum_{t=0}^{T-1} \nabla \log \pi(a_t|b_t;\theta)^T \cdot w + C. \tag{5.26}$$

To compute the parameters w, a least squares method can be used to solve the following set of equations

$$r_n = \sum_{t=0}^{T_n-1} \nabla \log \pi(a_t^n | b_t^n; \theta)^T \cdot w + C \qquad \forall n \in \{1, \ldots, N\}, \tag{5.27}$$

where r_n is the reward observed at the end of each dialogue. Substituting (5.26) into (5.25) gives:

$$\nabla R(\theta) \approx \nabla R(\theta; w) = \frac{1}{N} \sum_{n=1}^{N} \sum_{t=0}^{T_n-1} \nabla \log \pi(a_t^n | b_t^n; \theta) \nabla \log \pi(a_t^n | b_t^n; \theta)^T \cdot w. \tag{5.28}$$

Note that in (5.28), the baseline B was cancelled by the constant C and it can be shown that C is an optimal constant baseline which minimizes the variance of the gradient [30]. Once the gradient estimate (5.28) is computed, it can be used to update the policy parameters by: $\theta' \leftarrow \theta + \beta \nabla R(\theta; w)$, where β determines the step size.

5.5.1 *Natural Actor Critic Algorithm*

Although (5.28) can be used to optimise the policy parameters, the use of the "plain" gradient yields rather poor convergence properties since methods using this gradient often suffer from extremely flat plateaus in the expected reward function. In contrast, the *natural gradient* defined as

$$\widetilde{\nabla} R(\theta) = F^{-1}(\theta) \nabla R(\theta), \tag{5.29}$$

where $F(\theta)$ is the Fisher Information Matrix, does not suffer from such behaviour [1]. Based on this idea, a family of natural actor critic (NAC) algorithms which estimates a natural gradient of the expected reward function has been developed [19]. An appealing feature of these algorithms is that in practice, the Fisher Information matrix does not need to be explicitly computed. Inspecting (5.28), it can be observed that the expression

$$\frac{1}{N} \sum_{n=1}^{N} \sum_{t=0}^{T_n-1} \nabla \log \pi(a_t^n | b_t^n; \theta) \nabla \log \pi(a_t^n | b_t^n; \theta)^T \tag{5.30}$$

is in fact an estimate of the Fisher Information matrix. Thus, (5.28) can be written as $\nabla R(\theta) \approx F(\theta) w$ and the natural gradient of the expected reward is simply

$$\widetilde{\nabla} R(\theta) = F^{-1}(\theta) \nabla R(\theta) \approx F^{-1}(\theta) F(\theta) w = w. \tag{5.31}$$

Hence, the weights which minimise the mean square error in the linear function approximation of $R(H)$ are in fact the components of the natural gradient. A fortunate consequence of this is that use of the natural gradient not only improves performance, but it also leads to a less computationally intensive algorithm.

Once the parameters w of the approximation of the reward function, $R(H;w)$, have been computed, the policy parameters θ can be iteratively improved by $\theta' \leftarrow \theta + \beta w$.

5.5.2 Natural Actor and Belief Critic Algorithm

In Sect. 5.4 it was shown that the expectation propagation algorithm can be used to infer not only the distribution over the unobserved dialogue states but also the dialogue model parameters. The main advantage of the EP algorithm is that since it is unsupervised, it does not need annotated data. However, EP does not guarantee a maximisation of the expected reward, whereas ideally, both the dialogue model and the policy parameters should be designed to maximise the expected reward (5.5). The rest of this section will describe the natural actor and belief critic (NABC) algorithm which offers a solution to this problem [14].

The NABC algorithm extends policy gradient methods so that the dialogue model parameters are optimised jointly with the policy parameters. Note that this algorithm does not need annotated data, although information about the rewards received in the sampled dialogues is still necessary.

The stochastic policy given in (5.3) can be written in more detail to show the dependency of the belief state b_t on the dialogue history h_t and the model parameters τ

$$\pi(a_t|b(\cdot|h_t;\tau);\theta) \approx \frac{\mathrm{e}^{\theta^T \cdot \phi_{a_t}(b(\cdot|h_t;\tau))}}{\sum_{\tilde{a}} \mathrm{e}^{\theta^T \cdot \phi_{\tilde{a}}(b(\cdot|h_t;\tau))}}. \tag{5.32}$$

The policy π is now clearly seen to depend on the parameters τ. The difficulty with using policy gradient methods for learning the parameters of the dialogue model $\mathcal{M}(\tau)$ is that since the function ϕ, which extracts features from the belief state, is usually a handcrafted function of non-continuous features, the policy is not usually differentiable with respect to τ. However, this problem can be alleviated by assuming that the model parameters τ come from a prior distribution $p(\tau;\alpha)$ that is differentiable with respect to the parameters α. Then, this prior can be sampled to give the model parameters during the estimation of the gradient of the expected reward.

The goal of NABC is therefore to learn the parameters α of the prior distribution for the model parameters τ together with the policy parameters θ while maximising the expected reward (5.5). Let the model parameters τ be sampled from the prior at the beginning of each dialogue, then τ becomes part of the observed history $h_t = \{\tau, a_0, o_1, \ldots, a_{t-1}, o_t\}$. Similarly the complete history H_t is extended for

τ to give $H_t = \{\tau, s_0, a_0, o_1, s_1, \ldots, a_{t-1}, o_t, s_t\}$. Given the formulation above, the expected reward (5.5) can be written as

$$R(\alpha,\theta) = \int p(H;\alpha,\theta)R(H)\mathrm{d}H \tag{5.33}$$

which depends on both θ and α and its gradient can be estimated by

$$\nabla R(\alpha,\theta) \approx \frac{1}{N}\sum_{n=1}^{N} \nabla \log p(H_n;\alpha,\theta)(R(H_n) - B), \tag{5.34}$$

where the probability of the trajectory H is defined as

$$p(H;\alpha,\theta)=p(s_0)p(\tau;\alpha)\prod_{t=1}^{T} p(o_t|s_t;\tau)p(s_t|a_{t-1},s_{t-1};\tau)\pi(a_{t-1}|b(\cdot|h_{t-1};\tau);\theta). \tag{5.35}$$

Consequently, the log-gradient $p(H;\alpha,\theta)$ has the following form

$$\nabla \log p(H;\alpha,\theta) = \left[\nabla_\alpha \log p(\tau;\alpha)^T, \sum_{t=0}^{T-1} \nabla_\theta \log \pi(a_t|b(\cdot|h_t;\tau);\theta)^T\right]^T. \tag{5.36}$$

Similar to the derivation of the policy gradient algorithms, this leads to solving the following set of equations using the least squares method

$$\begin{aligned} r_n = &\left[\nabla_\alpha \log p(\tau_n;\alpha)^T, \sum_{t=0}^{T_n-1} \nabla_\theta \log \pi(a_t^n|b(\cdot|h_t^n;\tau);\theta)^T\right] \\ &\times [w_\alpha^T, w_\theta^T]^T + C \quad \forall n \in \{1,\ldots,N\}, \end{aligned} \tag{5.37}$$

where r_n is the reward observed at the end of each dialogue and the solution $[w_\alpha^T, w_\theta^T]^T$ is the natural gradient of the expected reward.

Similar to the NAC algorithm, the $[w_\alpha^T, w_\theta^T]^T$ vector can be used to iteratively improve the policy parameters and the prior of the dialogue model parameters: $\theta' \leftarrow \theta + \beta_\theta w_\theta$, $\alpha' \leftarrow \alpha + \beta_\alpha w_\alpha$. Finally when the estimate of the parameters converges, the dialogue model parameters τ are computed as the expectation of the prior distribution $p(\tau;\alpha)$.

A specific form of the NABC algorithm which updates only the model parameters can also be derived. This method called Natural Belief Critic algorithm can be used not only with stochastic policies but also handcrafted policies [13, 14]. This is especially useful in commercial applications where the behaviour of dialogue systems is constrained by specific and very often handcrafted requirements (see Chap. 8).

5.5.3 The Dialogue Model Parameters Prior

In order to use NABC in practice, a prior for the model parameters τ is needed. Since the parameters of the Bayesian network described in Sect. 5.3 are parameters of multiple multinomial distributions, a product of Dirichlet distributions provides a convenient prior. Formally, for every node $j \in \{1, \dots, J\}$ in the Bayesian Network, there are parameters τ_j describing a probability $p(j|par(j); \tau_j)$ where the function $par(j)$ defines the parents of the node j. Let $|par(j)|$ be the number of distinct combinations of values of the parents of j. Then, τ_j is composed of the parameters of $|par(j)|$ multinomial distributions and it is structured as follows: $\tau_j = \left[\tau_{j,1}, \dots, \tau_{j,|par(j)|}\right]$. Consequently, a prior for τ_j can be formed from a product of Dirichlet distributions: $\prod_{k=1}^{|par(j)|} Dir(\tau_{j,k}; \alpha_{j,k})$, parameterised by $\alpha_{j,k}$. Let the vector $\tau = [\tau_1, \dots, \tau_J]$ be a vector of all parameters in the Bayesian network. Then, the probability $p(\tau; \alpha)$ from (5.37) can be defined as:

$$p(\tau; \alpha) = \prod_{j=1}^{J} \prod_{k=1}^{|par(j)|} Dir(\tau_{j,k}; \alpha_{j,k}), \tag{5.38}$$

for which a closed form solution of the log-gradient exists and this can be used in (5.37) to compute the natural gradient of the expected reward.

5.5.4 Evaluation

To give an indication of performance, an experimental evaluation of the NAC and NABC algorithms was conducted using the BUDS dialogue system as outlined in Sect. 5.3 [27]. The BUDS dialogue manager can use both a handcrafted policy and a stochastic policy of the form described in Sect. 5.2.

To compare performance of the NAC and NABC algorithms, three dialogue systems were built for the CAMINFO Restaurant domain described in Sect. 5.3:

- **HDC**—a system using a handcrafted dialogue policy and a set of finely tuned handcrafted dialogue model parameters
- **NAC**—a system using a stochastic policy optimised using the NAC algorithm and a set of handcrafted dialogue model parameters
- **NABC**—a system using a combined set of stochastic policy and dialogue model parameters jointly optimised using the NABC algorithm

The systems were trained and tested using an agenda-based user simulator [22] which incorporates a semantic concept confusion model to enable training and testing across a range of semantic error rates. The reward function awards -1 in each dialogue turn, and at the end of a dialogue, it awards 20 for a successful dialogue and 0 for an unsuccessful one. A dialogue is considered successful if a suitable

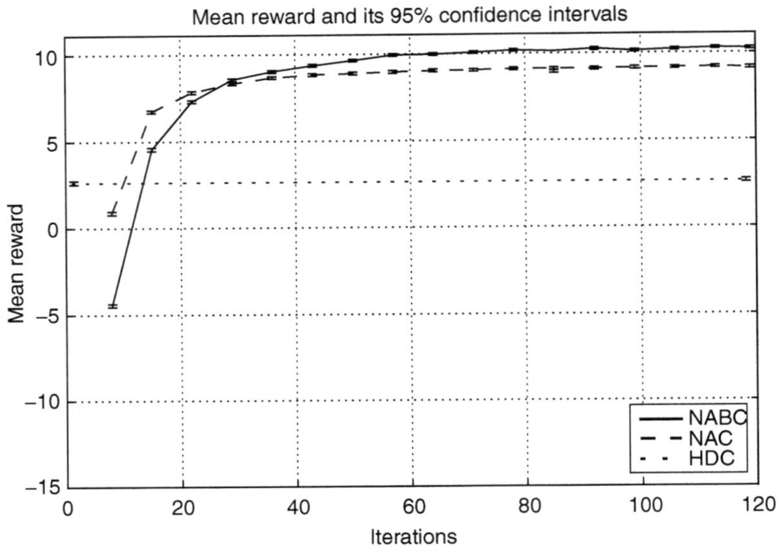

Fig. 5.4 Average reward across all error rates during training of the NAC and NABC systems. The horizontal line gives the performance of the handcrafted system (HDC)

venue is offered and all further pieces of information are given. In the case where no venue matches the constraints, the dialogue is deemed successful if the system tells the user that no venue matches and a suitable alternative is offered. Since a typical dialogue will require around five or six turns to complete, this implies that around 15 represents an upper bound on the achievable mean reward.

The systems were trained by executing 120 iterations with the simulator set to produce error rates between 0% and 50%, uniformly distributed among the dialogues. The total number of sampled dialogues per iteration was 32k. In total, 881 parameters were estimated for the policy $\mathcal{P}$ and 577 for the model $\mathcal{M}$. Both the policy parameters and the parameters of the prior of the dialogue model were initialised by uninformative (uniform) parameters.

Figure 5.4 compares the learning curves of the NAC and the NABC systems. In the beginning, the NAC algorithm learns faster as it does not have to learn the model parameters; however, as more iterations of training are completed, the performance of the fully trainable system outperforms the baseline with the handcrafted dialogue model parameters. After 120 iterations, both the model and the policy parameters converge.

The performance of the HDC, NAC, and NABC systems plotted as a function of the semantic error rate is depicted in Fig. 5.5. At each error rate, 5,000 dialogues were simulated and to reduce the variance of results, the training and evaluation procedures were executed 5 times and the results averaged. The results show that the NAC algorithm is very efficient in optimising the policy parameters. For example, at 35% error rate, the mean reward was increased from −1.96 to 7.35. The results also

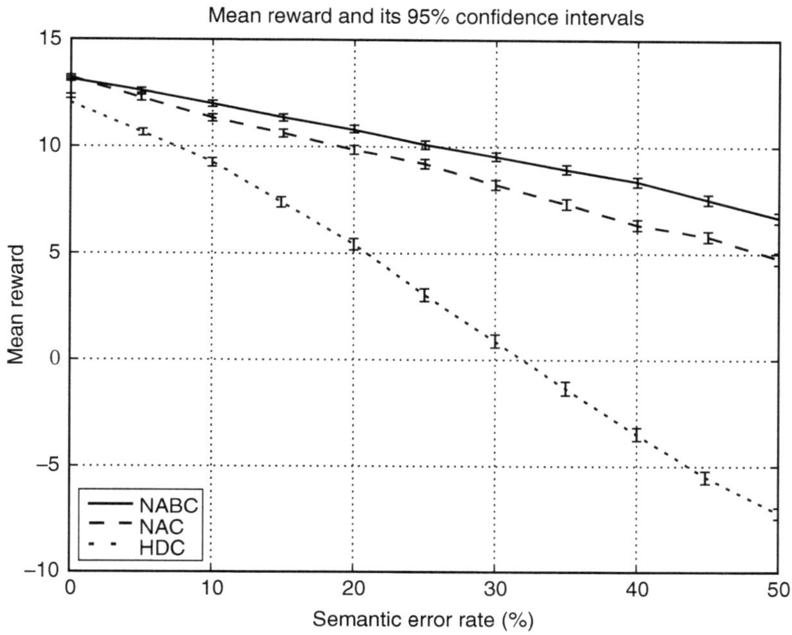

Fig. 5.5 Comparison of the HDC, NAC and NABC systems. The graph plots the mean reward against the semantic error rate

show that the additional optimization of the dialogue model parameters, performed by joint optimization of the policy and the dialogue model parameters using the NABC algorithm, significantly improves the performance. For example, at 35% error rate, the mean reward was improved by 23% from 7.35 to 9.03.

5.6 Gaussian Processes for Policy Modelling

As mentioned in the introduction, the aim of a dialogue policy is to map each belief state into an optimal action, i.e., the action that leads to the highest reward at the end of the dialogue. However, exact solutions to policy optimisation are only tractable for problems with very small state-action spaces. Parametric approximations, such as the softmax stochastic policy described in Sect. 5.5, offer tractable solutions but typically require the basis functions that extract the key features of the belief state to be handcrafted. Furthermore, gradient-based optimisation such as NAC requires large numbers of dialogues to form accurate estimates of the gradient, and even then, they can only be optimal subject to the approximation inherent in the selected basis functions.

Alternative methods of policy optimisation discretise the belief space into grid points. This allows the use of standard reinforcement learning algorithms to optimise the policy. However, in order to make real-world dialogue tasks tractable, the belief space must be compressed into a so-called summary space which leads to similar approximation issues. Furthermore, even when compressed to a few hundred grid points, around 10^5 dialogues are still required for optimisation [9]. This number is too large to allow training with real users so the interaction is normally performed with a simulated user. This raises further issues relating to both the cost and complexity of building an adequate simulator, and the degree to which such a simulator can truly reflect real human user behaviour (see Chap. 4).

In the light of the above, this section briefly describes a rather different approach to policy modelling and optimisation using Gaussian processes to provide non-parametric Bayesian models for function approximation. An advantage of Bayesian approaches is that they offer a principled way of incorporating prior knowledge about the underlying task into the learning process, which gives them the potential to improve learning rates. It is important that the dependencies in the different parts of the belief state space are taken into consideration during learning. Gaussian processes are able to incorporate the prior knowledge of these dependencies elegantly through the choice of a so-called *kernel function*, the purpose of which is to describe correlations in different parts of the space. In addition, Gaussian processes allow the variance of the posterior to be estimated, thus modelling the uncertainty of the approximation. This is particularly useful for dialogue management, since for every belief state-action pair the Gaussian process not only provides a policy estimate, but it also provides a measure of the uncertainty inherent in taking that action in that belief state.

The next section explains how Gaussian processes can be used for policy modelling. In Sect. 5.6.2, the core idea of the method is explained on a "toy" dialogue problem, where different aspects of GP learning are examined and the results are compared. Section 5.6.3 then demonstrates how this methodology can be applied to a real-world dialogue problem.

5.6.1 Gaussian Process Reinforcement Learning

The role of a dialogue policy π is to map each belief state $\mathbf{b}$ into an action a so as to maximise the expected discounted cumulative reward, which is defined by the Q-function

$$Q(\mathbf{b},a) = \max_{\pi} \mathcal{E}_{\pi} \left\{ \sum_{i=t+1}^{T} \gamma^{i-t-1} r_i | b_t = \mathbf{b}, a_t = a \right\}, \tag{5.39}$$

where r_i is the reward obtained at turn i, T is the dialogue length and γ is the discount factor, $0 < \gamma \leq 1$.

The problem of obtaining the optimal policy is thus equivalent to the problem of obtaining the optimal Q-function:

$$\pi(\mathbf{b}) = \arg\max_{a} Q(\mathbf{b},a). \tag{5.40}$$

In this case, both the policy and its associated Q-function are deterministic. However, both can be modelled as stochastic processes. Firstly, the Q-function can be modelled as a zero mean Gaussian process by providing a kernel function which defines the correlations in different parts of the belief state and action spaces:

$$Q(\mathbf{b},a) \sim \mathcal{GP}\left(0, k((\mathbf{b},a),(\mathbf{b},a))\right). \tag{5.41}$$

The kernel $k(\cdot,\cdot)$ is often factored into separate kernels over the belief state and action spaces $k_{\mathcal{B}}(\mathbf{b},\mathbf{b})k_{\mathcal{A}}(a,a)$.

Given this prior and some observed belief state–action pairs, the posterior of the Q-function can be computed. For a sequence of belief state–action pairs $\mathbf{B}_t = [(\mathbf{b}^0,a^0),\ldots,(\mathbf{b}^t,a^t)]^\mathsf{T}$ visited in a dialogue and the corresponding observed immediate rewards $\mathbf{r}_t = [r^1,\ldots,r^t]^\mathsf{T}$, the posterior of the Q-function for any belief state-action pair $(\mathbf{b},a)$ is defined by

$$Q(\mathbf{b},a)|\mathbf{r}_t,\mathbf{B}_t \sim \mathcal{N}(\overline{Q}(\mathbf{b},a), cov((\mathbf{b},a),(\mathbf{b},a))), \tag{5.42}$$

where

$$\begin{aligned}
\overline{Q}(\mathbf{b},a) &= \mathbf{k}_t(\mathbf{b},a)^\mathsf{T}\mathbf{H}_t^\mathsf{T}(\mathbf{H}_t\mathbf{K}_t\mathbf{H}_t^\mathsf{T}+\sigma^2\mathbf{H}_t\mathbf{H}_t^\mathsf{T})^{-1}\mathbf{r}_t,\\
cov((\mathbf{b},a),(\mathbf{b},a)) &= k((\mathbf{b},a),(\mathbf{b},a))\\
&\quad -\mathbf{k}_t(\mathbf{b},a)^\mathsf{T}\mathbf{H}_t^\mathsf{T}(\mathbf{H}_t\mathbf{K}_t\mathbf{H}_t^\mathsf{T}+\sigma^2\mathbf{H}_t\mathbf{H}_t^\mathsf{T})^{-1}\mathbf{H}_t\mathbf{k}_t(\mathbf{b},a),\\
\mathbf{H}_t &= \begin{bmatrix} 1 & -\gamma & \cdots & 0 & 0\\ 0 & 1 & \cdots & 0 & 0\\ \vdots & \ddots & \ddots & \vdots & \vdots\\ 0 & \cdots & 0 & 1 & -\gamma \end{bmatrix},\\
\mathbf{K}_t &= [\mathbf{k}_t((\mathbf{b}^0,a^0)),\ldots,\mathbf{k}_t((\mathbf{b}^t,a^t))],\\
\mathbf{k}_t(\mathbf{b},a) &= [k((\mathbf{b}^0,a^0),(\mathbf{b},a)),\ldots,k((\mathbf{b}^t,a^t),(\mathbf{b},a))]^\mathsf{T}
\end{aligned} \tag{5.43}$$

and where σ^2 is a measure of the assumed additive noise in the estimate of the Q-function and γ is the discount factor. Matrix $\mathbf{H}_t$ captures the discounting of the reward in Eq. 5.39 (see [6, 12] for details). Matrix $\mathbf{K}_t$, also called the *Gram matrix*, defines the correlations between the data points.

The marginal likelihood of the observed rewards has an analytic solution

$$\mathbf{r}_t|\mathbf{B}_t \sim \mathcal{N}(\mathbf{0},\mathbf{H}_t(\mathbf{K}_t+\sigma^2\mathbf{I})\mathbf{H}_t^\mathsf{T}), \tag{5.44}$$

and this provides the relation between the observed rewards and the kernel function via the Gram matrix.

The kernel function can be parameterised in the form $k(\cdot,\cdot) = k(\cdot,\cdot;\Theta)$ where Θ are the kernel parameters, also called the *hyper-parameters*. Note that these parameters are different to the policy parameters described earlier in this chapter.

The main difference is that the way these parameters are set does not restrict the optimality of the solution; instead, their effect is mainly on the number of data points needed and the resulting accuracy of the optimal solution. If the kernel function is parameterised, the Gram matrix is also parameterised $\mathbf{K}_t = \mathbf{K}_t(\Theta)$. Given a corpus of visited belief states, with the actions taken and resulting rewards, the hyper-parameters can be estimated by maximising the marginal likelihood from (5.44).

Due to the matrix inversion in (5.42), the computational complexity of calculating the Q-function posterior is $O(t^3)$, where t is the number of data points. In the case of a dialogue system, the number of points used for estimation will be equal to the total number of turns, summed over all dialogues, and this number can be very large. For this reason, a sparse approximation is needed which can ensure that all the data points are taken into account whilst reducing the computational complexity. One such method is the kernel span sparsification method [7]. The basic idea of this method is to select a small subset of data points, the *representative points*, with which the kernel function can be effectively approximated. It can be shown that this reduces the complexity of calculating the posterior to $O(tm^2)$, where m is the number of representative points.

The description so far has only given the stochastic model for the Q-function. However, what is required for the dialogue management is the model for the policy π. One way of defining a policy is via an ε-greedy approach. It requires setting of an additional parameter ε which balances how often the action is taken based on the current best estimate of the Q-function mean—the exploration—and how often an action is taken randomly—the exploration. Thus, the model becomes

$$a = \begin{cases} \arg\max_a \overline{Q}(\mathbf{b}, a) & \text{with probability } 1-\varepsilon \\ \text{random} & \text{with probability } \varepsilon. \end{cases} \tag{5.45}$$

Alternatively, active learning may be incorporated into the action selection process in order to provide efficient data selection [4, 16]. The main idea behind active learning is to select only the data points that contribute the most to the estimate. Here, similar to [5], active learning can be used for more efficient exploration. During exploration, actions are chosen based on the variance of the GP estimate for the Q-function, and during exploitation, actions are chosen based on the mean:

$$a = \begin{cases} \arg\max_a \overline{Q}(\mathbf{b}, a) & \text{with probability } 1-\varepsilon \\ \arg\max_a cov\left((\mathbf{b}, a), (\mathbf{b}, a)\right) & \text{with probability } \varepsilon. \end{cases} \tag{5.46}$$

Both of these approaches require a manual balancing of exploration and exploitation by tuning the parameter ε. Since the Gaussian process for the Q-function defines a Gaussian distribution for every belief state–action pair (Eq. 5.42), when a new belief point $\mathbf{b}$ is encountered, for each action $a_i \in \mathcal{A}$, there is a Gaussian distribution $Q(\mathbf{b}, a_i) \sim \mathcal{N}(\overline{Q}(\mathbf{b}, a_i), cov((\mathbf{b}, a_i), (\mathbf{b}, a_i))))$. Sampling from these Gaussian distributions gives a set of Q-values for each action $\{Q^i(\mathbf{b},a_i)\}$ from which the action with the highest sampled Q-value can be selected:

$$a = \arg\max_{a_i} Q^i(\mathbf{b},a_i). \tag{5.47}$$

Thus, this approach maps the GP approximation of the Q-function into a stochastic policy model.

All of these approaches allow observations to be processed sequentially, in direct interaction with the user, whether real or simulated. A specific and commonly used algorithm for achieving this is GP-Sarsa which is particularly suited to online episodic reinforcement learning [6, 7, 12].

5.6.2 Gaussian Process Reinforcement Learning for a Simple Voice Mail Dialogue Task

The practical application of the above ideas to a spoken dialogue task and the use of the GP-Sarsa algorithm will first be illustrated using a very simple "toy" voice mail dialogue system [29]. This system has just three states: the user asked for the message either to be saved or deleted, or the dialogue ended; and three actions: ask the user what to do, save or delete the message. The observation of what the user actually wants to do at each turn is corrupted with noise. For both learning and evaluation, a simulated user is used which generates an observation error with probability 0.3 and terminates the dialogue after at most 10 turns. In the final state, the system receives a reward of 10 for operating as intended or a penalty of -100 otherwise, for example because it deleted rather than saving the mail. Each intermediate state receives the penalty of -1. In order to keep the problem simple, a model defining the transition and observation probabilities is assumed so that the belief can be easily updated, but policy optimisation is performed on-line.

The kernel function must accurately represent prior knowledge about the Q-function correlations and it must be defined for both states and actions. Since the action space is discrete, a simple δ kernel can be defined over actions

$$k(a,a') = 1 - \delta_a(a'), \tag{5.48}$$

where δ_a is the Kronecker delta function.

In contrast, the state space is continuous space, and it is assumed that points in belief space which are in some sense similar will be more strongly correlated. Here, four different kernel functions are investigated, listed in Table 5.1. Each kernel function defines a different correlation. The polynomial kernel views elements of the state vector as features, the dot product of which defines the correlation. They can be given different relevance r_i in the parameterised version. The Gaussian kernel accounts for smoothness, i.e., if two states are close to each other, the Q-function in these states is correlated. The scaled norm kernel defines positive correlations in

Table 5.1 Kernel functions

Kernel function	Expression
Polynomial	$k(\mathbf{x},\mathbf{x}') = \langle \mathbf{x},\mathbf{x}' \rangle$
Parameterised poly.	$k(\mathbf{x},\mathbf{x}') = \sum_{i=1}^{D} \frac{x_i x_i'}{r_i^2}$
Gaussian	$k(\mathbf{x},\mathbf{x}') = p^2 \exp \frac{-\lVert\mathbf{x}-\mathbf{x}'\rVert^2}{2\sigma_k^2}$
Scaled norm	$k(\mathbf{x},\mathbf{x}') = 1 - \frac{\lVert\mathbf{x}-\mathbf{x}'\rVert^2}{\lVert\mathbf{x}\rVert^2\lVert\mathbf{x}'\rVert^2}$

the points that are close to each other and a negative correlation otherwise. This is particularly useful for the voice mail problem, where, if two belief states are very different, taking the same action in these states generates a negatively correlated reward.

Since this toy problem is very simple, it is possible to compute an exact solution for the optimal policy, for example, using the POMDP solver toolkit [3]. Hence, the policy learnt by GP-Sarsa can be compared with the exact solution, and in order to assess the efficiency with which it learns, it can also be compared with a standard grid-based algorithm such as the Monte Carlo control (MCC) algorithm [24, 34].

The dialogue manager was therefore trained in interaction with the simulated user, and the performance was compared between the grid-based MCC algorithm and GP-Sarsa using the different kernel functions from Table 5.1. Kernel hyper-parameters were optimised using 300 sample dialogues obtained using the exact optimal policy. All the algorithms use an ε-greedy approach where the exploration rate ε was fixed at 0.1. In order to reduce the affects of statistical variation, for every training set-up, exactly the same training iterations were performed using 1,000 different random generator seedings. After every 20 dialogues, the resulting 1,000 partially optimised policies were evaluated by testing it on 1,000 dialogues and averaging the results.

The results are shown in Fig. 5.6. As can be seen, the grid-Based MCC algorithm has a relatively slow convergence rate. GP-Sarsa with the polynomial kernel exhibited a learning rate similar to MCC in the first 300 training dialogues, continuing with a more upward learning trend. The parameterised polynomial kernel performs slightly better. The Gaussian kernel, however, achieves a much faster learning rate. The scaled norm kernel achieved close to optimal performance in 400 dialogues, with a much higher convergence rate then the other methods. Thus, GP-Sarsa can learn close to optimal policies much more quickly than standard reinforcement learning algorithms, but the choice of kernel is clearly very important.

5.6.3 Gaussian Process Reinforcement Learning for a Real-World Tourist Information Task

To show that the GP approach can be scaled to practical dialogue systems, this section will briefly describe its application to policy optimisation in the Cambridge

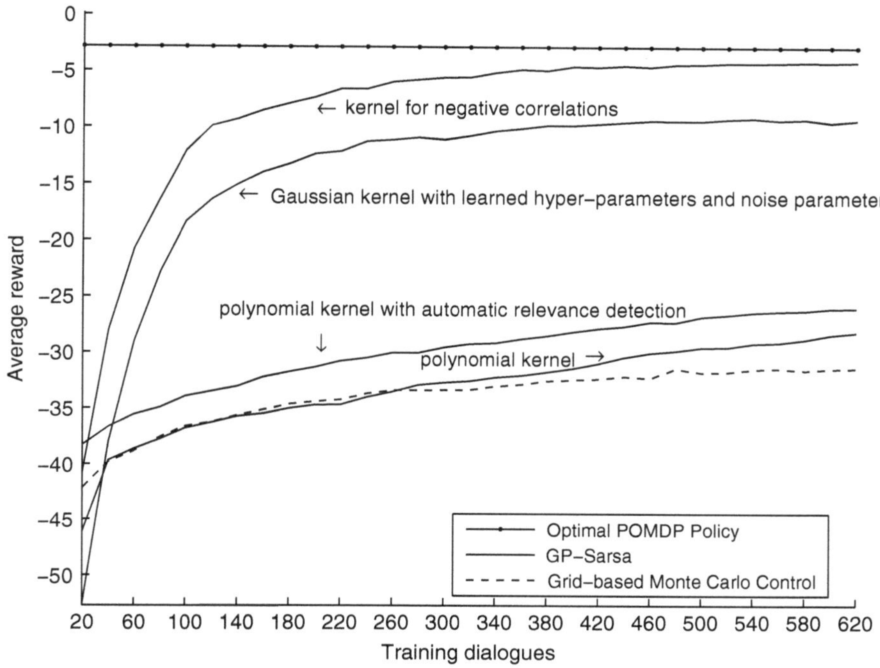

Fig. 5.6 Evaluation results on voice mail task

CamInfo tourist information system which provides information about restaurants, bars, hotels and other tourist attractions in the Cambridge area. The database consists of more than 400 entities each of which has up to 10 attributes that the user can query.

The CamInfo system can be configured to run with either the BUDS-based dialogue manager described in Sect. 5.3 or with the hidden information state (HIS) dialogue manager [34]. Since the HIS system is slightly easier to adapt to GP-Sarsa, that version will be described here.

The summary state in the HIS system is a four-dimensional space consisting of two elements that are continuous (the probability of the top two hypotheses) and two discrete elements (one relating the portion of the database entries that matches the top partition and the other relating to the last user action type). The summary action space is discrete and consists of 11 elements. The nature of the HIS state space is quite different from that of the toy problem, and kernels that have negative correlations, such as the scaled norm kernel, cannot be used. Instead, and somewhat contrary to the results for the toy problem, a second order polynomial kernel $k(\mathbf{x},\mathbf{x}') = (\langle \mathbf{x},\mathbf{x}' \rangle + 1)^2$ turns out to be the most appropriate for the continuous elements (see [12]). For the discrete elements, the δ-kernel (5.48) is used.

As previously, policy optimisation is performed by interacting with a simulated user, and the system receives a reward of 20 or 0, depending on whether or not the

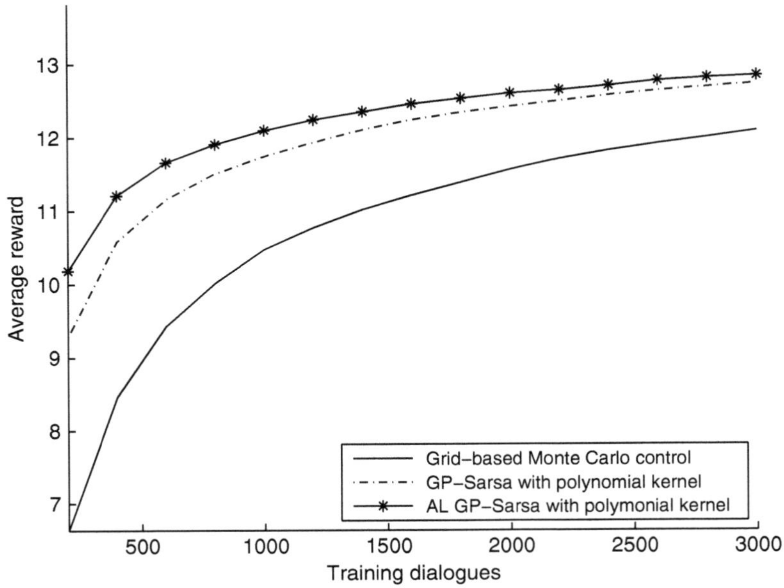

Fig. 5.7 Evaluation results on CAMINFO task

dialogue was successful, less the number of turns taken to fulfil the user's request. Figure 5.7 shows a graph of performance as a function of the number of dialogues used for policy optimisation for both GP-Sarsa and the baseline grid-based MCC algorithm.

The results show that in the very early stage of learning, i.e., during the first 400 dialogues, the GP-based method learns much faster. Furthermore, the learning process can be accelerated by using active learning (AL) where the actions are selected based on the estimated uncertainty as in (5.46). After performing many iterations, both the GP-Sarsa and the grid-based MCC algorithms converge to the same performance.

Figure 5.7 shows that active learning using an ε-greedy policy with variance exploration learns faster than a standard random exploration. However, this graph only displays the average reward. If GP is to be used to learn online with real users, then the variance of the reward is also a concern.

Figure 5.8 compares the learning rates of an ε-greedy policy with variance exploration with that of the fully stochastic policy given by (5.47). As can be seen, the learning rates are similar, but the variance of the stochastic policy is much reduced. This suggests that for practical systems, the use of the stochastic policy is preferable because it reduces the risk of taking bad actions during learning. When real users are involved, this benefit may well be significant.

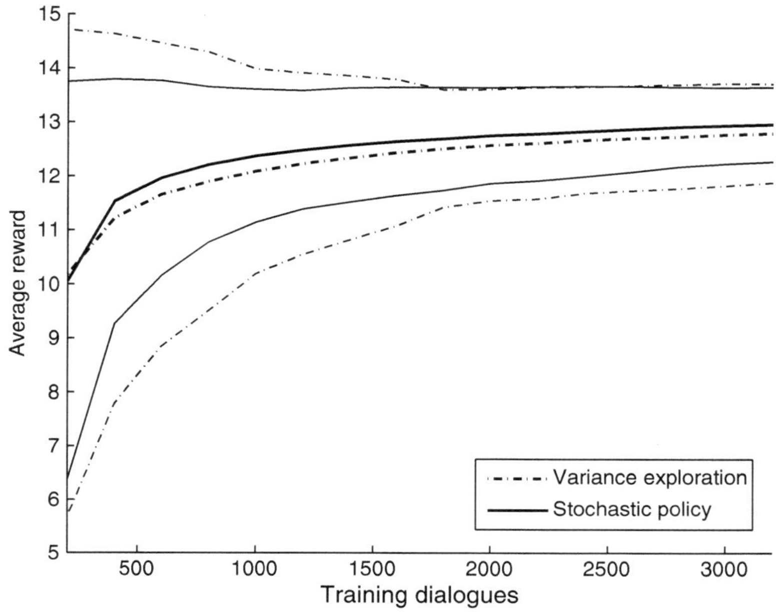

Fig. 5.8 Average reward vs. training epoch with a simulated user for a stochastic policy compared to an ε-greedy policy with variance exploration. The upper and lower bounds in each case denote 95% confidence intervals

5.7 Conclusions

Statistical SDS based on the framework of POMDPs offer considerable potential for reducing costs and increasing robustness. However, practical systems have complex probability models and very large state spaces. The availability of efficient algorithms for optimising model parameters and policies is therefore essential for developing these systems further.

This chapter has described three different approaches to optimisation. Firstly, it was shown that by incorporating the model parameters into a Bayesian network representation of the dialogue model and using an extended form of inference algorithm called expectation-propagation (EP), it is possible to learn model parameters on-line in parallel with belief monitoring. Secondly, it was shown that standard policy gradient methods are not only effective for policy exploration, they can also be used to optimise model parameters. This is particularly appealing because unlike EP, the gradient approach directly maximises the expected reward. Finally, the use of a non-parametric approach was presented in which the Q function is modelled as a Gaussian process. The key features of this approach are that it includes an explicit model of state space correlations and it explicitly provides a measure of the uncertainty in the current Q function estimate. These allow the GP approach to achieve much faster learning and safer exploration of alternative actions.

The optimisation methods described in this chapter make important steps towards building more flexible SDS which can learn from data and which are robust to speech recognition errors. Increasing the speed of learning and extending the range of learning to include the model parameters as well as the policy makes it possible to optimise and adapt systems on-line in direct interaction with human users. As well as allowing more accurate training and avoiding the cost of building expensive user simulators, these developments open the possibility of building significantly more powerful systems. Firstly, they will be able to track and adapt their behaviour to individual user preferences. Secondly, they will able to learn the structure of the Bayesian network directly from data, allowing them to acquire new domain knowledge as it is used and to evolve eventually over time towards truly robust and natural conversational interfaces.

References

1. Amari, S.: Natural gradient works efficiently in learning. Neural. Comput., **10**(2), 251–276 (1998)
2. Bishop, C.: Pattern Recognition and Machine Learning, Springer (2006)
3. Cassandra, A.R.: POMDP solver [Computer Software] (2005). Available from http://www.cassandra.org/pomdp/code/
4. Cohn, D., Atlas, L., Ladner, R.: Improving Generalization with Active Learning. Mach. Learn. **15**, 201–221 (1994)
5. Deisenroth, M.P., Rasmussen, C.E., Peters, J.: Gaussian Process Dynamic Programming. Neurocomputing **72**(7-9), 1508–1524 (2009)
6. Engel, Y.: *Algorithms and Representations for Reinforcement Learning*. PhD thesis, Hebrew University (2005)
7. Engel, Y., Mannor, S., Meir, R.: Reinforcement learning with Gaussian processes. In: Proceedings of ICML (2005)
8. Engel, E., Mannor, S., Meir, R.: Reinforcement learning with Gaussian processes. In: ICML 2005, Bonn, Germany (2005)
9. Gašić, M., Keizer, S., Mairesse, F., Schatzmann, J., Thomson, B., Yu, K., Young, S.: Training and evaluation of the HIS-POMDP dialogue system in noise. In: Proceedings of SIGDIAL (2008)
10. Gašić, M., Jurčíček, F., Keizer, S., Mairesse, F., Thomson, B., Yu, K., Young, S.J.: Gaussian Processes for Fast Policy Optimisation of a POMDP Dialogue Manager for a Real-world Task. In: SigDial 2010, Tokyo, Japan (2010)
11. Gašić, M., Young, S.: Effective Handling of Dialogue State in the Hidden Information State POMDP Dialogue Manager. ACM Transactions on Speech and Language Processing **7**(3), 2011
12. Gašić, M.: Statistical dialogue modelling. PhD thesis, University of Cambridge (2011)
13. Jurčíček, F., Thomson, B., Keizer, S., Mairesse, F., Gašić, M., Yu, K., Young, S.J.: Natural Belief-Critic: a reinforcement algorithm for parameter estimation in statistical spoken dialogue systems. In Takao Kobayashi, Keikichi Hirose, and Satoshi Nakamura, editors, Proc. Interspeech, 90–93, ISCA (2010)
14. Jurčíček, F., Thomson, B., Young, S.J.: Natural Actor and Belief Critic: Reinforcement algorithm for learning parameters of dialogue systems modelled as POMDPs. ACM Transactions on Speech and Language Processing **7**(3), 2011

15. Keizer, S., Gašić, M., Mairesse, F., Thomson, B., Yu, K., Young, S.: Modelling user behaviour in the HIS-POMDP dialogue manager. In: Proceedings of SLT, pp. 121–124, 2008
16. MacKay, D.J.C.: Information-based objective functions for active data selection. Neural. Comput. **4**(4), 590–604 (1992)
17. Minka, T.P.: Expectation Propagation for Approximate Bayesian Inference. In: Proc 17th Conf in Uncertainty in Artificial Intelligence, pp. 362–369. Seattle, Morgan-Kaufmann (2001)
18. Paquet, U.: *Bayesian inference for latent variable models*. PhD thesis, University of Cambridge (2007)
19. Peters, J., Schaal, S.: Natural Actor-Critic. m Neurocomputing **71**(7-9), 1180–1190 (2008)
20. Rasmussen, C.E., Williams, C.K.I.: Gaussian Processes for Machine Learning. MIT Press (2006)
21. Roy, N., Pineau, J., Thrun, S: Spoken Dialogue Management Using Probabilistic Reasoning. In: Proceedings of the ACL 2000, 2000
22. Schatzmann, J., Stuttle, M.N., Weilhammer, K., Young, S.: Effects of the user model on simulation-based learning of dialogue strategies. In: IEEE ASRU '05: Proc. IEEE Workshop Automatic Speech Recognition and Understanding (2005)
23. Schatzmann, J.: *Statistical User and Error Modelling for Spoken Dialogue Systems*. PhD thesis, University of Cambridge (2008)
24. Sutton, R.S., Barto, A.G.: *Reinforcement Learning: An Introduction*. Adaptive Computation and Machine Learning. MIT Press, Cambridge, Massachusetts (1998)
25. Thomson, B.: *Statistical methods for spoken dialogue management*. PhD thesis, University of Cambridge (2009)
26. Thomson, B., Schatzmann, J., Young, S.J.: Bayesian Update of Dialogue State for Robust Dialogue Systems. In: Int Conf Acoustics Speech and Signal Processing ICASSP, Las Vegas (2008)
27. Thomson, B., Young, S.J.: Bayesian update of dialogue state: A POMDP framework for spoken dialogue systems. Computer Speech and Language **24**(4), 562–588 (2010)
28. Thomson, B., Jurčíček, F., Gašić, M., Keizer, S., Mairesse, F., Yu, K., Young, S.J.: Parameter learning for POMDP spoken dialogue models. In: IEEE Workshop on Spoken Language Technology (SLT 2010), Berkeley, CA (2010)
29. Williams, J.D.: *Partially Observable Markov Decision Processes for Spoken Dialogue Management*. PhD thesis, University of Cambridge (2006)
30. Williams, R.J.: Simple statistical gradient-following algorithms for connectionist reinforcement learning. Mach. Learn. **8**, 229–256 (1992)
31. Williams, J.D., Poupart, P., Young, S.: Factored partially observable Markov decision processes for dialogue management. In: Proceedings of the IJCAI Workshop on Knowledge and Reasoning in Practical Dialog Systems, 2005
32. Williams, J.D., Young, S.J.: Partially Observable Markov Decision Processes for Spoken Dialog Systems. Computer Speech and Language **21**(2), 393–422 (2007)
33. Young, S.J.: Talking to Machines (Statistically Speaking). In: Int Conf Spoken Language Processing, Denver, Colorado (2002)
34. Young, S.J., Gašić, M., Keizer, S., Mairesse, F., Schatzmann, J., Thomson, B., Yu, K.: The Hidden Information State Model: a practical framework for POMDP-based spoken dialogue management. Computer Speech and Language **24**(2), 150–174 (2010)

Chapter 6
Statistical Approaches to Adaptive Natural Language Generation

Oliver Lemon, Srini Janarthanam, and Verena Rieser

Employing statistical models of users, generation contexts and of natural languages themselves has several potentially beneficial features: the ability to train models on real data, the availability of precise mathematical methods for optimisation, and the capacity to adapt robustly to previously unseen situations.

This chapter will describe recent statistical approaches to generating natural language in spoken dialogue systems (SDS), covering several methods developed in the CLASSIC project: Reinforcement Learning (RL) approaches as developed by [22, 25, 33, 35, 44] and language generation using graphical models and active learning with data collected through crowd-sourcing [36]. The former approach has been applied to higher-level decisions in natural language generation (NLG) such as content and attribute selection [33, 35, 44, 45], as well as some low-level decisions such as lexical choice [22, 25], while the latter focusses on data-driven surface realisation. (See [50] and Fig. 6.1 for a representation of the NLG decision structure for SDS.)

In this chapter, the RL approaches are presented in Sects. 6.3 and 6.4, and work using graphical models is discussed in Sect. 6.5. We also discuss recent work which jointly optimises higher-level and lower-level NLG decisions, using combinations of these types of approaches [12, 13], in Sect. 6.6.

The reinforcement learning approach presented below shares many features with more traditional planning approaches to NLG, but it uses statistical machine learning models to develop adaptive NLG components for SDS. Rather than emulating human behaviour in generation (which can be sub-optimal), these methods can even find strategies for NLG which improve upon human performance. Some very encouraging test results have recently been obtained with real users of systems developed using these methods, which we will survey here. For example, in an evaluation with real users (see Sect. 6.3.5), a trained information presentation

O. Lemon (✉) • S. Janarthanam • V. Rieser
Heriot-Watt University, Edinburgh, EH14 4AS, UK
e-mail: o.lemon@hw.ac.uk; sc445@hw.ac.uk; v.t.rieser@hw.ac.uk

O. Lemon and O. Pietquin (eds.), *Data-Driven Methods for Adaptive Spoken Dialogue Systems*, DOI 10.1007/978-1-4614-4803-7_6,

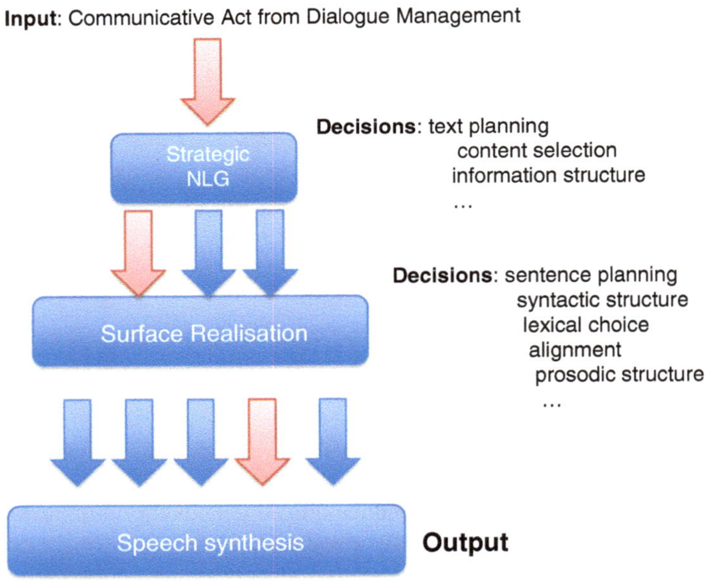

Fig. 6.1 NLG decisions in a SDS

strategy significantly improved dialogue task completion, with up to a 9.7 % increase compared to a deployed dialogue system which used conventional, hand-coded presentation prompts.

Turning to surface realisation, using factored language models, a fully data-driven statistical language generator was developed to produce utterances in a tourist information domain (see Sect. 6.5). Its output was shown not to differ significantly from human paraphrases (over 200 test utterances). Furthermore, active learning from semantically labelled utterances collected through crowd-sourcing was shown to significantly improve performance on sparse training sets.

This chapter will explain the motivations behind these approaches, and related work, and will describe several case studies which illustrate their advantages, with reference to recent empirical results in the areas of information presentation and referring expression generation (REG). This presentation will include discussion of user simulations for training NLG components (Sect. 6.4.2), reward functions for training NLG, and evaluation methods and metrics (Sect. 6.3.5). Finally, we provide a critical outlook for future work in this direction, in Sect. 6.7.

6.1 Motivation: The Task for NLG Systems in SDS

NLG is often characterised as choosing "how to say" something once "what to say" has been determined. For example, depending on the specific domain of an interactive system (in this example a restaurant-search system), the dialogue act `inform(price=cheap)` might be uttered as *"This restaurant is economically priced"* or *"That's a cheap place to eat"*, and the act `present_results (item3,item4,item7)` could be uttered as, *"There are 3 places you might like. The first is Bonsai on Richmond street, the second is Sushiya on Dalry Road, and the third is Kanpai on Grindlay Street"*, or else perhaps by comparing the three restaurants by contrasting their features. The central issue for NLG for SDS is that the context (linguistic or extra-linguistic) within which generation takes place is *dynamic*. This means that different expressions might need to be generated at different moments, even to convey the same concept or refer to the same object for the same user. Thus, good NLG systems must appropriately *adapt* their output to a changing context. This requires modelling the interaction context and determining policies and realisation strategies that are sensitive to transitions in this context.

In principle, NLG in a SDS comprises a wide variety of decisions, ranging over content structuring, choice of referring expressions, use of ellipsis, aggregation and choice of syntactic structure, to the choice of intonation markers for synthesised speech. In computational dialogue systems, "what to say" is usually determined by a dialogue manager (DM) component, via planning, hand-coded rules, finite state machines, or learned policies, and "how to say it" is then very often defined by simple templates or hand-coded rules which define appropriate word strings to be sent to a speech synthesiser or screen.

Machine learning approaches to NLG could in theory have several key advantages over template-based and rule-based approaches (we discuss other approaches to "trainable" NLG in Sect. 6.2) in dialogue systems:

- The ability to adapt to fine-grained changes in dialogue context.
- A data-driven development cycle.
- Provably optimal action policies with a precise mathematical model for action selection.
- The ability to generalise to unseen dialogue states.

In the remainder of this chapter, we will present previous work on adaptive NLG (Sect. 6.2), and then case studies in the areas of information presentation (Sect. 6.3) and REG (Sect. 6.4). We maintain a particular focus on results of recent evaluations with real users. We conclude with an outlook for this research area in Sect. 6.7.

6.2 Previous Work in Adaptive NLG

There have been a variety of approaches to generating natural language from data, though most of them have focused on static, non-interactive settings, such as summarising numerical time-series data [42], and the majority have generated text meant for human consumption via reading rather than for listening to via synthesised speech in an interactive dialogue system. Few systems have considered generating for different types of users or in changing contexts.

One of the most sophisticated adaptive approaches in recent years generated different types of spoken information presentations for different types of users (in the flight information domain) [37]. This system determined theme/rheme information structure (including contrasts for different user preferences) and employed word-level emphasis for spoken output. It deployed a rule-based planning system for content structuring based on predefined user models and a combinatory categorial grammar (CCG) realiser component [57]. However, this system, and other user-adaptive systems such as EPICURE [10], TAILOR [39] and M-PIRO [19], depends on a static user model, which the system is given before performing generation. A new challenge is generation based on uncertain and dynamic models of the user and their context. These models need to be estimated during interaction with the system and cannot be preloaded [25].

In addition, the great majority of currently deployed NLG modules within interactive systems use only template-based generation (fixed patterns of output where some phrases are variables filled at runtime). Here, the key decisions of "what to say and how to say it" are still being made by nonadaptive template-based systems, which always say the same things in the same way, no matter who the user is or where they are in space, time or social context. Recent advances have shown that statistical planning methods outperform traditional template-based and rule-based adaptive approaches (e.g. [22, 25, 43, 44]).

The overall framework of NLG as planning under uncertainty was first set out in 2008 [34] (and later developed in [33, 44]), where each NLG action is a sequential decision point, based on the current dialogue/generation context and the expected long-term utility or "reward" of the action. (For an accessible guide to the background concepts of Markov decision processes (MDPs) and reinforcement learning, see [51].) Note that this approach has recently been adopted by a variety of other researchers [12, 13]. Other recent approaches describe this task as planning, e.g. [30, 31], or as contextual decision making according to a cost function [53], but not as a statistical planning problem, where uncertainty in the stochastic environment is explicitly modelled.

Rule-based and supervised learning approaches have both been proposed for learning and adapting dynamically *to different users* during interactions. Such systems learn from a user at the start of an interaction and then later adapt to the domain knowledge of the user. However, they either require expensive expert knowledge resources to hand-code the inference rules for adaptation [5] or else a large corpus of expert-layperson interactions from which adaptive strategies can be learned and modelled, using methods such as Bayesian networks [1]. RL approaches learn in the absence of these expensive resources, and crowdsourcing can be used to collect useful corpora quite rapidly [36]. It is also not clear how supervised approaches choose between when to seek more information from the user and when to adapt to them. In the study below, we show how to learn such a decision automatically, in Sect. 6.4.

We now examine case studies which illustrate some of the advantages of the RL approach: information presentation (in Sect. 6.3) and REG (in Sect. 6.4) in SDS.

6.3 Adaptive Information Presentation

In information-seeking dialogues, the information presentation (IP) phase is the primary contributor to dialogue duration [54] and is therefore a central aspect of system design. During this phase, the system returns a set of items (or "hits") from a database, which match the user's current search constraints. The system's IP strategy should support the user in choosing between the available options, and ultimately in selecting the most suitable option by structuring and selecting information. A central problem in this task is the trade-off between presenting "enough" information to the user (e.g. helping them to feel confident that they have a good overview of the search results) versus keeping the utterances short and understandable, especially when using spoken output. This trade-off will be represented in the reward function for learning (Sect. 6.3.3).

Rieser et al. [43] showed that IP for SDS can be treated as a data-driven joint optimisation problem, and that this outperforms a supervised model of human "wizard" behaviour on a particular IP task (presenting sets of search results to a user). A similar approach has been applied to the problem of REG in dialogue [22, 28].

The information presentation problem contains many decisions available for exploration. For instance, which presentation strategy to apply (*NLG strategy selection*), how many attributes of each item to present (*attribute selection*), how to rank the items and attributes according to different models of user preferences (*attribute ordering*), how many (specific) items to tell them about (*coverage*), how many sentences to use when doing so (*syntactic planning*) and which words to use (*lexical choice*). All these parameters (and potentially many more) can be varied and, ideally, jointly optimised based on user judgements and preferences.

Broadly speaking, previous research on IP for SDS has been concerned with two major problems: (1) IP strategy selection and (2) content or attribute selection. Prior work has presented a variety of **IP strategies** for structuring information (see examples in Table 6.1). For example, the SUMMARY strategy is used to guide the user's "focus of attention". It draws the user's attention to relevant attributes by grouping the current results from the database into clusters, e.g. [11, 40]. Other studies investigate a COMPARE strategy, e.g. [38, 55], while most work in SDS uses a RECOMMEND strategy, e.g. [61].

A proof-of-concept study [44] showed that each of these strategies has its own strengths and drawbacks, dependent on the particular context in which information needs to be presented to a user, while [43] explores possible combinations of the strategies, e.g. SUMMARY followed by RECOMMEND, e.g. [58], see Fig. 6.2.

Prior work on **content or attribute selection** has used a "summarise and refine" approach [6, 40, 41]. This method employs utility-based attribute selection with respect to how each attribute (e.g. price or food type in restaurant search) of a set of items helps to narrow down the user's goal to a single item. Related work explores a user modelling approach, where attributes are ranked according to user preferences [11, 59]. Our data collection and training environment incorporate these approaches.

Table 6.1 Example realisations, generated when the user provided `cuisine=Indian`, and where the wizard has also selected the additional attribute `price` for presentation to the user

Strategy	Example utterance
SUMMARY no UM	I found 26 restaurants, which have Indian cuisine. Eleven of the restaurants are in the expensive price range. Furthermore, ten of the restaurants are in the cheap price range and five of the restaurants are in the moderate price range
SUMMARY UM	Twenty six restaurants meet your query. There are ten restaurants which serve Indian food and are in the cheap price range. There are also 16 others which are more expensive
COMPARE by item	The restaurant called Kebab Mahal is an Indian restaurant. It is in the cheap price range. And the restaurant called Saffrani, which is also an Indian restaurant, is in the moderate price range
COMPARE by attribute	The restaurant called Kebab Mahal and the restaurant called Saffrani are both Indian restaurants. However, Kebab Mahal is in the cheap price range while Saffrani is moderately priced
RECOMMEND	The restaurant called Kebab Mahal has the best overall quality amongst the matching restaurants. It is an Indian restaurant, and it is in the cheap price range

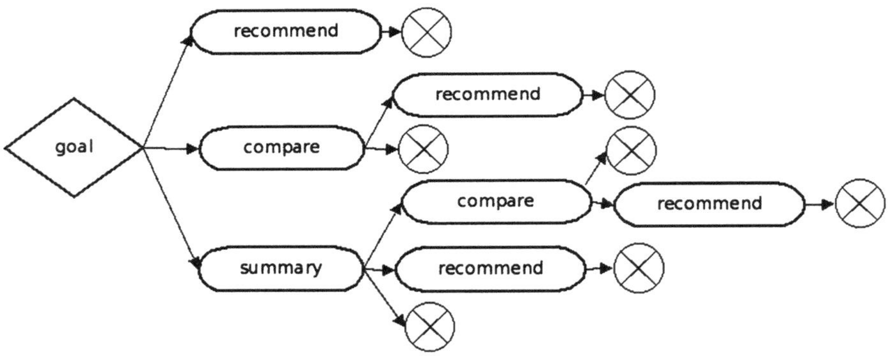

Fig. 6.2 Possible information presentation structures (X=stop generation)

Rieser at al. [43] were the first to apply a data-driven method to this whole decision space (i.e. combinations of information presentation strategies as well as attribute selection) and to show the utility of both lower-level features (e.g. from the surface realiser) and higher-level features (i.e. information from dialogue management) for this problem. Previous work only focused on individual aspects of the problem (e.g. how many attributes to generate, or when to use a SUMMARY), using a pipeline model for SDS with DM features as input, and where NLG has no knowledge of lower-level features (e.g. behaviour of the surface realiser). Rieser at al. [43] used reinforcement learning (RL) [51] as a statistical planning framework to explore the contextual features for making these decisions, and proposed a joint optimisation method for IP strategies combining content structuring and attribute selection.

Here, the example task was to present a set of search results (e.g. restaurants) to users. In particular, considering seven possible policies for structuring the content (see Fig. 6.2): recommending one single item, comparing two items, summarising all of them or ordered combinations of those actions, e.g. first summarise all the retrieved items and then recommend one of them. The IP module has to decide which action to take next, how many attributes to mention and when to stop generating. A sentence generator based on the stochastic sentence planner SPaRKy [50] was employed for surface generation.

In particular, the following IP strategies were realised (see examples in Table 6.1):

- SUMMARY of all matching restaurants with or without a user model (UM), following [40]. The approach using a UM assumes that the user has certain preferences (e.g. cheap) and only tells him about the relevant items, whereas the approach with no UM lists all the matching items.
- COMPARE the top two restaurants by item (i.e. listing all the attributes for the first item and then for the other) or by attribute (i.e. directly comparing the different attribute values).
- RECOMMEND the top-ranking restaurant (according to UM).

6.3.1 Reinforcement Learning: Model and Training

We formulate the problem of information presentation as a MDP, where states are NLG dialogue contexts and actions are NLG decisions. Each state–action pair is associated with a transition probability, which is the probability of moving from state s at time t to state s' at time $t+1$ after having performed action a when in state s. This transition probability is computed by the environment model (i.e. the user simulation and realiser, see Sect. 6.3.2), and explicitly captures the uncertainty in the generation environment. This is a major difference compared to other non-statistical planning approaches. Each transition is also associated with a reinforcement signal (or "reward") r_{t+1} describing how good the result of action a was when performed in state s. The aim of the MDP is to maximise long-term expected reward of its decisions, resulting in a *policy* which maps each possible state to an appropriate action in that state.

IP was treated as a *hierarchical* joint optimisation problem, where first one of the IP structures (1–3) is chosen and then the number of attributes is decided, as shown in Fig. 6.3. At each generation step, the MDP can choose 1–5 attributes (e.g. cuisine, price range, location, food quality, and/or service quality). Generation stops as soon as the user is predicted to select an item, i.e. the IP task is successful. (Note that the same constraint is operational for the WoZ baseline system.)

States are represented as sets of context features. The state space comprises "lower-level" features about the realiser behaviour (two discrete features representing the number of attributes and sentences generated so far) and three binary features

Fig. 6.3 State-action space for the IP problem

$$
\left[\begin{array}{ll}
\text{ACTION:} & \left[\textit{IP:} \left\{ \begin{array}{l} \text{SUMMARY} \\ \text{COMPARE} \\ \text{RECOMMEND} \end{array} \right\} \left\{ \texttt{attr: 1-5} \right\} \right] \\
\text{STATE:} & \left[\begin{array}{l}
\texttt{attributes: } \{\texttt{1-15}\} \\
\texttt{sentence: } \{\texttt{2-18}\} \\
\texttt{dbHitsFocus: } \{\texttt{1-100}\} \\
\texttt{userSelect: } \{\texttt{0,1}\} \\
\texttt{userAddInfo: } \{\texttt{0,1}\} \\
\texttt{userElse: } \{\texttt{0,1}\}
\end{array} \right]
\end{array}\right]
$$

representing the user's predicted next action, as well as "high-level" features provided by the DM (e.g. current database hits in the user's focus (`dbHitsFocus`)). The policy was trained for 60,000 iterations, using the SHARSHA algorithm [48] with linear function approximation [51], and the simulation environment described in Sect. 6.3.2.

Note here that the learner is optimising with respect to uncertain user reactions but also because the behaviour of the surface realiser is unknown (it is part of the training environment).

6.3.2 User Simulations for the IP Problem

User simulations are commonly used to train strategies for Dialogue Management; see Chap. 4. A user simulation for NLG is very similar, in that it is a predictive model of the most likely next user act.[1] However, this predicted user act does not actually change the overall dialogue state (e.g. by filling slots), but it only changes the generator state. In other words, the NLG user simulation tells us what the user is most likely to do next, *if we were to stop generating now.*

In information-seeking systems we are particularly interested in the following user reactions:

1. `select`: The user chooses one of the presented items, e.g. *"Yes, I'll take that one"*. This reply type indicates that the information presentation was sufficient for the user to make a choice.

[1] Similar to the internal user models applied in recent work on POMDP (partially observable markov decision process) dialogue managers [16, 18, 61] for estimation of user act probabilities.

2. `addInfo`: The user provides more attributes, e.g. *"I want something cheap"*. This reply type indicates that the user has more specific requests, which s/he wants to specify after being presented with the current information.
3. `requestMoreInfo`: The user asks for more information, e.g. *"Can you recommend me one?"*, *"What is the price range of the last item?"*. This reply type indicates that the system failed to present the information the user was looking for.
4. `askRepeat`: The user asks the system to repeat the same message again, e.g. *"Can you repeat?"*. This reply type indicates that the utterance was either too long or confusing for the user to remember, or the TTS quality was not good enough, or both.
5. `silence`: The user does not say anything. In this case it is up to the system to take initiative.
6. `hangup`: The user closes the interaction.

We built user simulations using n-gram models of system (s) and user (u) acts, as first introduced by [15]. In order to account for data sparsity, different *discounting* ("smoothing") techniques were applied including *back-off*, using the CMU Statistical Language Modelling toolkit [8]. A **bi-gram** model[2] was built for the users' reactions to the system's IP structure decisions ($P(a_{u,t}|IP_{s,t})$), as well as a **tri-gram** (i.e. IP structure + attribute choice) model for predicting user reactions to the system's combined IP structure and attribute selection decisions: $P(a_{u,t}|IP_{s,t}, attributes_{s,t})$.

The performance of these models was evaluated by measuring dialogue similarity to the original data, based on the Kullback–Leibler (KL) divergence, as also used by, e.g. [9,27,29]. Raw probabilities as observed in the data were compared with the probabilities generated by the n-gram models using different discounting techniques for each context, see [43]. All the models had a small divergence from the original data (especially the bi-gram model), suggesting that they are reasonable simulations for training and testing NLG policies. We used the most similar user models for system training, and the most dissimilar user models for testing NLG policies, in order to test whether the learned policies are robust and adaptive to unseen dialogue contexts.

6.3.3 Data-Driven Reward Function

The reward/evaluation function was constructed from the Wizard-of-Oz (WoZ) data, using a stepwise linear regression, following the PARADISE framework [56]. This model selects the features which significantly influenced the users'

---

[2]Where $a_{u,t}$ is the predicted next user action at time t, $IP_{s,t}$ was the system's information presentation action at t, and $attributes_{s,t}$ are the attributes selected by the system at t.

ratings for the NLG strategy in the WoZ questionnaire. A value was also assigned to the user's reactions ($valueUserReaction$), similar to optimising task success for DM [61]. This reflects the fact that good IP strategies should help the user to `select` an item ($valueUserReaction = +100$) or provide more constraints `addInfo` ($valueUserReaction = \pm 0$), but the user should not do anything else ($valueUserReaction = -100$).

The regression in Eq. (6.1) ($R^2 = 0.26$) indicates that users' ratings are influenced by higher-level and lower-level features: Users prefer a focus on a small set of database hits (where $\#DBhits$ ranges over [1–100]), which will enable them to choose an item ($valueUserReaction$), while keeping the IP utterances short (where $\#sentence$ is in the range [2–18]). Database hits approximate the user's cognitive load by modelling the user's "focus of attention" dependent on the number of database hits presented by the last IP action, see [43].

$$\begin{aligned} Reward = {} & (-1.2) \times \#DBhits \\ & + (.121) \times valueUserReaction \\ & - (1.43) \times \#sentence. \end{aligned} \tag{6.1}$$

Note that the worst possible reward for an NLG move is therefore $(-1.20 \times 100) - (0.121 \times 100) - (18 \times 1.43) = -157.84$. This would be achieved by presenting 100 items to the user in 18 sentences[3], in such a way that the user ends the conversation unsuccessfully. The top possible reward is achieved in the rare cases where the system can immediately present one item to the user using just two sentences, and the user then selects that item, i.e. Reward = $-(1.20 \times 1) + (0.121 \times 100) - (2 \times 1.43) = 8.06$.

6.3.4 Results: Evaluation with Simulated Users

The learned strategies were then compared against the *WoZ baseline*. This baseline is constructed using supervised learning in order to mimic patterns observed in the original data, see [43].

For attribute selection, a majority baseline (randomly choosing between 3 and 4 attributes) was used, since the attribute selection models learned by supervised learning on the WoZ data did not show any significant improvement.

For training, the user simulation model most similar to the data was employed; see Sect. 6.3.2. For testing, the *different* user simulation model (the one which is most dissimilar to the data) was used.

We first investigated how well IP structure (without attribute choice) can be learned in increasingly complex generation **scenarios**. A generation scenario is a

---

[3]Note that the maximum possible number of sentences generated by the realiser is 18 for the full IP sequence SUMMARY+COMPARE+RECOMMEND using all the attributes.

Table 6.2 Test results for 1,000 simulated dialogues

Scenario	Wizard baseline average reward	RL average reward	RL (%) − baseline (%) = (%) improvement
1.1	−15.82(±15.53)	−9.90***(±15.38)	89.2 − 85.6 = 3.6
1.2	−19.83(±17.59)	−12.83***(±16.88)	87.4 − 83.2 = 4.2
2.1	−12.53(±16.31)	−6.03***(±11.89)	91.5 − 87.6 = 3.9
2.2	−14.15(±16.60)	−14.18(±18.04)	86.6 − 86.6 = 0.0
2.3	−17.43(±15.87)	−9.66***(±14.44)	89.3 − 84.6 = 4.7
2.4	−19.59(±17.75)	−12.78***(±15.83)	87.4 − 83.3 = 4.1

where *** denotes that the RL policy is significantly ($p < 0.001$) better than the baseline policy

combination of a particular kind of surface realiser (template vs. stochastic) along with different levels of variation introduced by certain features of the dialogue context. In general, the stochastic realiser introduces more variation in lower level features than the template-based realiser. The focus model introduces more variation with respect to #DBhits and #attributes as described in [43].We therefore investigated the following cases:

1.1. IP structure choice, template realiser: Predicted next user action varies according to the bi-gram model: $P(a_{u,t}|IP_{s,t})$; number of sentences and attributes per IP strategy are set by defaults, reflecting the use of a template-based realiser.

1.2. IP structure choice, stochastic realiser: The number of attributes per NLG turn is given at the beginning of each episode (e.g. set by the DM); sentences are generated according to the SPaRKy stochastic realiser model.

We then investigated different scenarios for *jointly* optimising IP structure (IPS) and attribute selection (Attr) decisions.

2.1. IPS+Attr choice, template realiser: Predicted next user action varies according to tri-gram model: $P(a_{u,t}|IP_{s,t}, attributes_{s,t})$; number of sentences per IP structure is set by default.

2.2. IPS+Attr choice, template realiser+focus model: Tri-gram user simulation with template realiser and focus of attention model with respect to #DBhits and #attributes.

2.3. IPS+Attr choice, stochastic realiser: Tri-gram user simulation with sentence/attribute relationship according to stochastic realiser.

2.4. IPS+Attr choice, stochastic realiser+focus: Predicted next user action varies according to tri-gram model+ focus of attention model + sentence/attribute relationship according to stochastic realiser.

The average final reward (see Eq. 6.1) gained by the baseline against the trained RL policies was compared in the different scenarios for each of 1,000 test runs, using a paired samples t-test. The results are shown in Table 6.2. In 5 out of 6 scenarios, the RL policy significantly ($p < 0.001$) outperformed the supervised learning ("wizard") baseline. Table 6.2 also reports the percentage of the top possible reward gained by the individual policies (RL % and baseline %), and the raw percentage improvement of the RL policy (% improvement). Note that the best possible (100%) reward can only be achieved in rare cases.

The learned RL policies show that lower-level features are important in gaining significant improvements over the baseline. The more complex the scenario, the harder it is to gain higher rewards for the policies in general (as more variation is introduced), but the relative improvement in rewards also increases with complexity: the baseline does not adapt well to the variations in lower-level features, whereas RL learns to adapt to the more challenging scenarios. [4]

For example, the RL policy for scenario 1.1 learned to start with a SUMMARY if the initial number of items returned from the database is high (>30). It will then stop generating if the user is predicted to select an item. Otherwise, it continues with a RECOMMEND. If the number of database items is low, it will start with a COMPARE and then continue with a RECOMMEND, unless the user selects an item.

In addition, the RL policy for scenario 1.2 learns to adapt to a more complex scenario: the number of attributes requested by the DM and produced by the stochastic sentence realiser. It learns to generate the whole sequence (SUMMARY+COMPARE+RECOMMEND) if #attributes is low (<3), because the overall generated utterance (final #sentences) is still relatively short. Otherwise the policy is similar to the one for scenario 1.1.

The RL policies for jointly optimising IP strategy and attribute selection learn to select the number of attributes according to the different generation scenarios 2.1–2.4. For example, the RL policy learned for scenario 2.1 generates a RECOMMEND with 5 attributes if the database hits are low (<13). Otherwise, it will start with a SUMMARY using 2 attributes. If the user is predicted to narrow down their focus after the SUMMARY, the policy continues with a COMPARE using 1 attribute only, otherwise attempts to help the user by presenting 4 attributes. It then continues with RECOMMEND (5), and stops as soon as the user is predicted to select one item.

The WoZ baseline policy generates all the possible IP structures (with 3 or 4 attributes) but is restricted to using only "high-level" features (see [43] for a detailed description of the WoZ model). Nevertheless, this wizard policy achieves up to 87.6 % of the possible reward on this task and so can be considered a serious baseline against which to measure performance. By beating this baseline, we show the importance of the "lower-level" features for optimising NLG.

The only case (scenario 2.2) where RL does not improve significantly over the baseline is where lower-level features do not play an important role for learning good strategies: Scenario 2.2 is only sensitive to higher-level features (DBhits).

---

[4]Note that the baseline does reasonably well in scenarios with variation introduced by only higher-level features (e.g. scenario 2.2).

6.3.5 CLASSiC Project Evaluation with Real Users

The NLG policy described above was then deployed in an extensive online user study, involving 131 users and more than 800 test dialogues, which explores its contribution to overall "global" task success in a SDS (see Chap. 7 for a discussion of evaluation methodologies).

The IP policy was integrated into the "CamInfo" system, a SDS developed in the CLASSiC project providing tourist information for Cambridge [62]. This baseline system was made accessible by phone using VoIP technology, enabling out-of-lab evaluation with large numbers of users. The speech recogniser (ASR), semantic parser (SLU) and dialogue manager (DM) were all developed at Cambridge University. For speech synthesis (TTS), the Baratinoo synthesiser, developed at France Telecom, was used.

The DM uses a POMDP (partially observable markov decision process) framework (see Chap. 5), allowing it to process N-best lists of ASR hypotheses and keep track of multiple dialogue state hypotheses. The DM policy is trained to select system dialogue acts given a probability distribution over possible dialogue states. It has been shown that such dialogue managers can exploit the information in the N-best lists (as opposed to only using the top ASR hypothesis) and are therefore particularly effective in noisy conditions [62].

The NLG component of this baseline system is a standard rule-based surface realiser covering the full range of system dialogue acts that the dialogue manager can produce. It has only one IP strategy, i.e. the system only provides information about database entries in the form of single venue recommendations (the RECOMMEND strategy, see Table 6.1). The attributes of the venue to be presented are selected heuristically. In the extended version of the system, the IP strategy is replaced by the trained NLG component, which is optimised to decide between different IP actions as described above.

The subjects were directed to a webpage with detailed instructions and for each task, a phone number to call and the scenario to follow. A scenario describes a place to eat in town, with some constraints, e.g. *"You want to find a moderately priced restaurant and it should be in the Riverside area. You want to know the address, phone number, and type of food,"*. After the dialogue, the subjects were asked to fill in a short questionnaire.

For the objective evaluation of the two systems task completion rates were measured. A full presentation and discussion of the results is given in [45]. In summary, we found that the trained information presentation strategy significantly improves dialogue task completion for real users, with up to a 9.7 % increase (30 % relative) compared to the deployed dialogue system which uses conventional, hand-coded presentation prompts. This result establishes the benefits of this overall methodology for building NLG systems.

6.4 Referring Expression Generation: Adapting to Unknown Users

A policy for REG allows a system to choose appropriate surface expressions to refer to domain entities in a dialogue setting. For instance, in a technical support conversation, an interactive system could choose to use more technical terms with an expert user, or to use more descriptive and general expressions with novice users, and a mix of the two with intermediate users of various sorts (see examples in Table 6.3). Here, we survey the model, method, and results from another evaluation with real users, for a reinforcement learning approach to user-adaptive REG policies, also using data-driven user simulations [25].

In human–human conversations dialogue partners learn about each other and adapt their language to suit their domain expertise [20]. This kind of adaptation is called `alignment through audience design` [3, 7]. We assume that a SDS must be capable of observing the user's dialogue behaviour, modelling his/her domain knowledge and adapting accordingly, just as human dialogue partners do. Therefore, unlike previous systems that use static user models, we investigated building a system which has to dynamically model the user's domain knowledge in order to adapt during conversations.

This work presents a corpus-driven framework for learning a user-adaptive REG policy from a small corpus of non-adaptive human–machine interactions and shows that a learned policy performs better than a simple hand-coded adaptive policy in terms of accuracy of adaptation, dialogue time and task completion rate when evaluated with real users.

Sections 6.4.1 and 6.4.2 describe the overall dialogue system framework and the user simulation model, respectively. Model training is presented in Sect. 6.4.3, and Sect. 6.4.4 presents the evaluation for different REG policies with real users.

6.4.1 Dialogue Manager and Generation Modules

The dialogue system presents the user with instructions to set up a broadband connection at home. In the WoZ experimental set up the system and the user interact using speech. However, in the RL set-up, they interact at the abstract level of dialogue actions and referring expressions. The system's objective is to learn to choose the appropriate referring expressions to refer to the domain entities in the instructions.

Dialogue Manager

The dialogue manager identifies the next dialogue act ($A_{s,t}$ where t denotes turn, s denotes system) to give to the user based on the dialogue management policy π_{dm}. Here, the dialogue management policy is fixed and is coded in the form

Table 6.3 Referring expression examples for two entities (from the corpus)

Jargon:	Please plug one end of the `broadband cable` into the `broadband filter`.
Descriptive:	Please plug one end of the `thin white cable with grey ends` into the `small white box`.

of a finite state machine. In this dialogue task, the system provides instructions for the user to either observe or manipulate the environment. When users ask for clarifications regarding referring expressions, the system clarifies (*provide_clar*) by giving information to enable the user to associate the expression with the intended referent. When the user responds in any other way, the instruction is simply repeated. The dialogue manager is also responsible for updating and managing the system state $S_{s,t}$. The system interacts with the user by passing both the system action $A_{s,t}$ and the referring expressions $REC_{s,t}$.

The Dialogue State and Dynamic User Model

The dialogue state $S_{s,t}$ represents the current state of the conversation. In addition to maintaining an overall dialogue state, the system maintains a user model $UM_{s,t}$ which records the initial domain knowledge of the user. This is a dynamic model which starts with a state where the system does not have any information about the user. Since the model is updated according to the user's behaviour, it may be inaccurate if the user's behaviour is itself uncertain. Hence, the user model is not always an accurate model of the user's knowledge and reflects a level of uncertainty about the user.

Each jargon referring expression `x` is represented by a three-valued variable in the dialogue state: `user_knows_x`. The three values that each variable takes are `yes`, `no` and `unknown`. The variables are updated using a simple user model update algorithm after the user's response each turn. Initially, each variable is set to `unknown`. If the user responds to an instruction containing the referring expression `x` with a clarification request, then `user_knows_x` is set to `no`. Similarly, if the user responds with appropriate information to the system's instruction, the dialogue manager sets `user_knows_x` is set to `yes`. Only the user's initial knowledge is recorded. This is based on the assumption that an estimate of the user's initial knowledge helps to predict the user's knowledge of the other referring expressions in the domain.

Referring Expression Generation Module

The REG module identifies the list of domain entities to be referred to and chooses the appropriate referring expression for each of the domain entities for each given dialogue act. In this study, we focussed only on the production of appropriate referring expressions to refer to domain entities mentioned in the dialogue act. The REG module chooses between the two types of referring expressions—jargon and descriptive. For example, the domain entity *broadband_filter* can be referred to using the jargon expression "broadband filter" or using the descriptive expression "small white box".[5] Although adaptation is the primary goal, it should be noted that in order to get information on the user that the system is dealing with, it needs to seek information by using jargon expressions.

The REG module operates in two modes—learning and evaluation. In the learning mode, the REG module is the learning agent. The REG module learns to associate dialogue states with optimal referring expressions. This is represented by a REG policy $\pi_{reg} : UM_{s,t} \rightarrow REC_{s,t}$, which maps the states of the dialogue (user model) to optimal referring expressions. The referring expression choices $REC_{s,t}$ is a set of pairs identifying the referent R and the type of expression T used in the current system utterance. For instance, the pair $(broadband\_filter, desc)$ represents the descriptive expression "small white box".

$$REC_{s,t} = \{(R_1, T_1), \ldots, (R_n, T_n)\}$$

In the evaluation mode, a trained REG policy interacts with unknown users. It consults the learned policy π_{reg} to choose the referring expressions based on the current user model.

6.4.2 User Simulations

This section presents user simulation models that simulate the dialogue behaviour of a real human user. Several user simulation models have been proposed for use in reinforcement learning of dialogue policies (see Chap. 4). However, they are suited only for learning dialogue management policies, and not NLG policies. In particular, the model presented below is the first to be sensitive to a system's choices of referring expressions.

Corpus-Driven Action Selection Model

The user simulation (US) receives the system action $A_{s,t}$ and its referring expression choices $REC_{s,t}$ at each turn. The US responds with a user action $A_{u,t}$ (u denoting user). This can either be a clarification request (cr) or an instruction response (ir). The US produces a clarification request cr based on the class of the referent $C(R_i)$, the type of the referring expression T_i, and the current domain knowledge of the user

[5]We will use italicised forms to represent the domain entities (e.g. *broadband_filter*) and double quotes to represent the referring expressions (e.g. "broadband filter").

for the referring expression $DK_{u,t}(R_i, T_i)$. Domain entities whose jargon expressions raised clarification requests in the corpus were listed, and those that had more than the mean number of clarification requests were classified as `difficult` and others as `easy` entities (e.g. *power_adaptor* is `easy`—all users understood this expression; *broadband_filter* is `difficult`). Clarification requests are produced using the following model.

$$P(A_{u,t} = cr(R_i, T_i)|C(R_i), T_i, DK_{u,t}(R_i, T_i))$$
$$\text{where}(R_i, T_i) \in REC_{s,t}.$$

One should note that the actual literal expression is not used in the transaction. Only the entity that it is referring to (R_i) and its type (T_i) are used. However, the above model simulates the process of interpreting and resolving the expression and identifying the domain entity of interest in the instruction. The user identification of the entity is signified when there is no clarification request produced (i.e. $A_{u,t} = none$). When no clarification request is produced, the environment action $EA_{u,t}$ is generated using the following model.

$$P(EA_{u,t}|A_{s,t}) \text{ if } A_{u,t}! = cr(R_i, T_i).$$

Finally, the user action is an instruction response which is determined by the system action $A_{s,t}$. Instruction responses can be either *provide_info*, *acknowledgement* or *other* based on the system's instruction.

$$P(A_{u,t} = ir|EA_{u,t}, A_{s,t}).$$

All the above models were trained on our corpus data using *maximum likelihood estimation* and smoothed using a variant of *Witten-Bell discounting*. The corpus contained dialogues between a non-adaptive dialogue system and real users. According to the data, clarification requests are much more likely when jargon expressions are used to refer to the referents that belong to the `difficult` class and which the user does not know about. When the system uses expressions that the user knows, the user generally responds to the instruction given by the system.

User Domain Knowledge

The user domain knowledge is initially set to one of several models at the start of every conversation. The models range from novices to experts which were identified from the corpus using `k-means` clustering (this partitions the users into clusters based on their distance from mean knowledge scores). A novice user knows only "power adaptor", whereas an expert knows all the jargon expressions, and intermediate users know some of them. It is assumed that users can interpret the descriptive expressions and resolve their references. Therefore, knowledge of descriptive expressions is not explicitly represented. Only the user's knowledge of jargon expressions is coded, using boolean variables representing whether the user knows the expression or not.

Corpus

The action selection model was trained on a small corpus of 12 non-adaptive dialogues between real users and a dialogue system. There were six dialogues in which users interacted with a system using just jargon expressions and six with a system using descriptive expressions. For more discussions of the user simulation models and the corpus, please refer to [21, 23, 26].

6.4.3 Training the REG Module

The REG module was trained (operated in learning mode) using the above simulations to learn REG policies that select referring expressions based on the user expertise in the domain. This section discusses how to code the learning agent's goals as reward, and then, how the reward function is used to train the learning agent.

Reward Function

A reward function was designed for the goal of adapting to each user's domain knowledge. The "adaptation accuracy" score (*AA*) calculates how accurately the REG agent chose the appropriate expressions for each referent *r*, with respect to the user's knowledge. So, when the user knows the jargon expression for *r*, the appropriate expression to use is jargon, and if s/he does not know the jargon, a descriptive expression is appropriate. Although the user's domain knowledge is dynamically changing due to learning, we base appropriateness on their initial state, because our objective is to adapt to the initial state of the user $DK_{u,initial}$. However, in reality, designers might want their system to account for users' changing knowledge as well. The accuracy per referent RA_r is calculated, followed by the overall mean adaptation accuracy (*AA*) over all referents as shown below:

$$RA_r = \frac{\#(appropriate\_expressions(r))}{\#(instances(r))},$$

$$Adaptation\ accuracy\ AA = \frac{1}{\#(r)}\Sigma_r RA_r.$$

Learning

The REG module was trained in learning mode using the above reward function using the SARSA reinforcement learning algorithm (with linear function approximation) [47, 51]. The training produced approx. 5,000 dialogues. The user simulation was calibrated to produce three types of users: novice, intermediate and expert, randomly but with equal probability.

Initially, during training, the REG policy chooses randomly between the referring expression types for each domain entity in the system utterance, irrespective of the user model state. Once the referring expressions are chosen, the system presents the user simulation with both the dialogue act and referring expression choices. The choice of referring expression affects the user's dialogue behaviour. For instance, choosing a jargon expression could evoke a clarification request from the user, based on which the dialogue manager updates the internal user model ($UM_{s,t}$) with the new information that the user is ignorant of the particular expression. It should be noted that using a jargon expression is an *information-seeking* move which enables the REG module to estimate the user's knowledge level. The same process is repeated for every dialogue instruction. At the end of the dialogue, the system is rewarded based on its choices of referring expressions. If the system chooses jargon expressions for novice users or descriptive expressions for expert users, penalties are incurred, and if the system chooses REs appropriately, the reward is high. On the one hand, those actions that fetch more reward are reinforced, and on the other hand, the agent tries out new state–action combinations to explore the possibility of greater rewards. Over time, it stops exploring new state–action combinations and exploits those actions that contribute to higher reward. The REG module learns to choose the appropriate referring expressions based on the user model in order to maximise the overall adaptation accuracy. Figure 6.4 shows how the agent learns using the data-driven simulation (**learned-DS**) during training.

6.4.4 Evaluation with Real Users

In order to compare the performance of the learned policy with a baseline, a simple rule-based policy was built. This baseline was chosen because it performed better in simulation, compared to a variety of other baselines [21]. It uses jargon for all referents by default and provides clarifications when requested. It exploits the user model in subsequent references after the user's knowledge of the expression has been set to either `yes` or `no`. Therefore, although it is a simple policy, it does adapt to a certain extent. We refer to this policy as the "jargon-adapt" policy. It should be noted that this policy was built in the absence of an expert domain knowledge and/or an expert-layperson corpus.

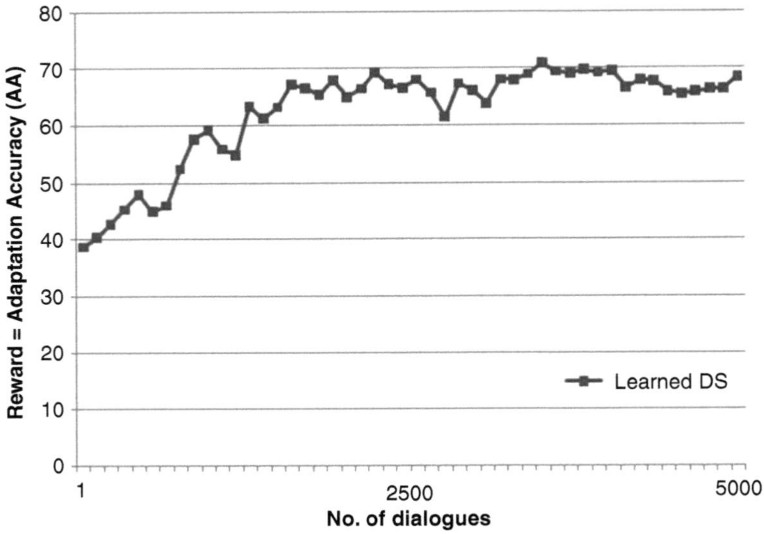

Fig. 6.4 Learning curve—training

The two policies were evaluated with real users: 36 university students from different backgrounds (e.g. Arts, Humanities, Medicine and Engineering) participated in the evaluation. Seventeen users were given a system with the jargon-adapt policy, and 19 users interacted with a system with the learned-DS policy. Each user was given a pre-task recognition test to record their initial domain knowledge. The experimenter read out a list of technical terms, and the users were asked to point out to the domain entities laid out in front of them. They were then given one of the two systems—learned or baseline—to interact with. Following the system instructions, they then attempted to set up the broadband connection. When the dialogue had ended, the users were given a post-task test where the recognition test was repeated and their responses were recorded. The user's broadband connection set-up was manually examined for task completion (i.e. the percentage of correct connections that they had made in their final set-up). The users were given the task completion results and were then given a user satisfaction questionnaire to evaluate the features of the system based on the conversation.

All users interacted with a system employing one of the two REG policies. The users' responses were intercepted by a human interpreter (or "wizard") and were annotated as dialogue acts, to which the automated dialogue manager responded with a system dialogue action (the dialogue policy was fixed). The wizards were not aware of the policy used by the system. The respective policies chose only the referring expressions to generate the system utterance for the given dialogue action. The system utterances were converted to speech by a speech synthesiser (Cereproc) and were played to the user.

Table 6.4 Evaluation with real users

Groups	JA	LDS
Strategy used	Jargon-adapt	Learned-DS
Adaptation accuracy (%)	63.91 (± 8.4)	84.72*** (± 4.72)
Dialogue time (mins)	7.86 (± 0.77)	6.98** (± 0.93)
Task completion rate (%)	84.7 (± 14.63)	99.47*** (± 2.29)
No. of turns	30.05 (± 2.33)	28.10* (± 1.85)
No. of clarification req	4.11 (± 2.36)	2.79 (± 1.43)

*Statistical significance ($p < 0.05$)
**Statistical significance ($p < 0.001$)
***Statistical significance ($p < 0.0001$)

Metrics

In addition to the adaptation accuracy metric mentioned above, other parameters from the conversation were measured in order to show how the learned adaptive policies compare with other policies on other dimensions. Dialogue time (DT) is the actual time taken for the user to complete the task. Task completion (TC) was measured by examining the user's broadband set-up after the task was completed (i.e. the percentage of correct physical connections that they had made in their final set-up).

Results

The performance of the two strategies for real users was compared using objective parameters and subjective feedback scores. Tests for statistical significance were performed using a Mann–Whitney test for two independent samples (due to non-parametric nature of the data).

Table 6.4 presents the mean accuracy of adaptation (AA), learning gain (LG), dialogue time (DT), and task completion (TC), produced by the two strategies. The learned-DS strategy produced more accurate adaptation than the jargon-adapt strategy ($p < 0.0001$). Higher accuracy of adaptation (AA) of the learned-DS strategy translates to less dialogue time ($p < 0.001$) and higher task completion ($p < 0.0001$) than the jargon-adapt policy. However, there was no significant difference in learning gain (LG).

Users' subjective feedbacks on different features of the systems were not very different from each other. However, users did notice that it was easier to identify domain objects with the learned-DS strategy than the jargon-adapt strategy ($p < 0.05$).

Analysis

The results show that the learned-DS strategy is significantly better than the hand-coded jargon-adapt policy in terms of adaptation accuracy, dialogue time and task completion rate. The initial knowledge of the users (mean pre-task recognition score) of the two groups was not significantly different from each other (JA $= 7.35 \pm 1.9$, LDS $= 6.57 \pm 2.29$). Hence, there is no bias on the user's pre-task score towards any strategy. While the learned-DS system adapts well to its users globally, the jargon-adapt system adapted only locally. This led to a higher task completion rate and lower dialogue times.

The learned-DS strategy learned to adapt to users in three ways. First, since the agent starts with no knowledge about the user, it learned to use jargon expressions as information-seeking moves at the start of a conversation. This behaviour elicits users' knowledge of the domain and populates the user model state. Second, it also learned the dependencies between domain entities and leveraged that knowledge to choose the appropriate referring expression for the user based on the user model. For example, a user who understands the expression "ethernet cable" might also understand "ethernet socket". On the other hand, if the user asked for clarification on the former, they would be presented with a descriptive expression for the latter. Third, it also learned to use jargon as an information-seeking move at appropriate moments in order to avoid being stuck in a local maximum. However, since it was trained to maximise the adaptation accuracy, the agent learned to restrict such moves and start predicting the user's domain knowledge as soon as possible. By learning to trade off between information-seeking and adaptation, the learned-DS policy produced a higher adaptation for real users with different domain knowledge levels.

In very recent work a similar approach has been applied to the generation of temporal referring expressions, such as *"The day after tomorrow"* or *"The same time on the 12th of June"* [28]. This work has been tested in field trials with real users.

6.5 Graphical Models and Active Learning for Surface Realisation

A further strand of NLG research in the CLASSiC project focussed on surface realisation, using factored language models. Here, a fully data-driven statistical language generator was developed to produce utterances in the tourist information domain [36]. This work developed the "BAGEL", system, which uses dynamic Bayesian networks to learn from semantically aligned data. This data was produced by untrained annotators using Amazon Mechanical Turk crowd-sourcing tools. First the annotators were asked to provide example utterances which matched abstract dialogue act specifications such as "Offer the venue Taj Mahal and provide the information type (restaurant), area (riverside), food (Indian), near (The Red Lion)", and they were then asked to align regions of their utterance with the semantic concepts in the dialogue act. There are therefore many valid ways of realising the same dialogue act, resulting in useful linguistic variation in the data. In addition, this data collection process is optimised through the use of certainty-based active learning. Here, the next dialogue act specification for annotation is chosen by the current model (i.e. the semantic inputs for which the model is least certain about its output realisation).

The BAGEL system then treats the generation task as search for the most likely sequence of realisation phrases given a (unordered) set of semantic stacks as input (see [36] for details of the stack-based semantic representation and the training process). Importantly, the system is able to generalise to semantic stacks which were not seen during training, because realisation phrases depend on underspecified stacks.

BAGEL was evaluated using human subjective ratings regarding naturalness and informativeness, also using Mechanical Turk. For models trained on 100 utterances or more, its output was shown not to differ significantly from human paraphrases (over 202 test cases). Furthermore, the active learning method was shown to significantly improve performance when training using limited amounts of data (both the training corpus and evaluation utterances are available at http://mi.eng.cam.ac.uk/~farm2/bagel).

6.6 Joint Optimization Approaches

The two strands of work discussed above focus on content structuring and attribute selection (the RL approach) and on surface realisation (BAGEL), but these processes are modelled separately and do not interact. However, as discussed earlier, such high-level and low-level decisions are in fact interdependent, and it is this interdependence that the work developed by [14] seeks to model.

Here, content selection, utterance planning and surface realisation decisions are optimised jointly using hierarchical reinforcement learning. This method is applied for the generation of navigation instructions in a virtual 3D world in the GIVE scenario [32]. Related work jointly optimises content selection and surface realisation using a log-linear classifier [2]. This recent work shows that jointly optimised policies outperform policies which have been optimised separately.

6.7 Research Directions

There are several directions in which this research can be developed. New research could improve over current methods by treating NLG as a hierarchy of statistical decision processes and could explore new application domains such as geo-spatial REG (e.g. for pedestrian route descriptions).

An interesting challenge for NLG, in general, is that of "generation under uncertainty" [24, 35], where language must be generated for users even though there is some uncertainty about their state. This uncertainty can be about their location, their gaze direction and their field of view, or even about their goals and preferences. Regarding generation under uncertainty, an interesting research direction will be to explicitly represent uncertainty about the generation context using techniques such as POMDPs, as presented in Chap. 5.

Future work could also include the predicted TTS quality [4] as a feature for optimising NLG decisions. Finally, the issue of incremental NLG in SDS must be tackled for more natural and efficient dialogue systems [46, 49]. Here, phenomena such as split utterances and barge-in are a research focus, and these are the target of statistical approaches to NLG in the EC FP7 PARLANCE project [60].

6.8 Conclusion

We have discussed recent work on data-driven methods for adaptive NLG in SDS, using two main case studies: information presentation (IP) and REG. At the time of writing, the reinforcement learning approach is being actively explored by a variety of researchers [12–14, 28, 45], and there is closely related work which also explores NLG as a process of maximising utility [17, 52].

For the IP problem a statistical optimisation framework was developed for content structure planning and attribute selection. This work was the first to apply a data-driven optimisation method to this decision space. An evaluation found that the trained information presentation strategy significantly improves dialogue task completion for real users, with up to a 9.7 % increase (30 % relative) compared to a deployed dialogue system which uses conventional, hand-coded presentation prompts.

This methodology provides new insights into the nature of the IP problem, which has previously been treated as a module following dialogue management with no access to lower-level context features.

In the second case study we showed that user-adaptive REG policies can also be learned using an RL framework and data-driven user simulations. This system learns to trade off between adaptive moves and information-seeking moves, to maximise the overall adaptation accuracy for a particular user's vocabulary. The learned policy started the conversation with information-seeking moves, learned a little about the user, and started adapting dynamically as the conversation progressed. The evaluation also showed that the learned policy performs better than a reasonable hand-coded policy with real users in terms of accuracy of adaptation, dialogue time, task completion and subjective evaluation.

It is also interesting to note that all the user studies show that an adaptive NLG component significantly contributes to the (perceived or objective) dialogue task success of the system (see Chap. 7). Thus, such data-driven adaptive NLG strategies have "global" effects on overall system performance. The data-driven planning methods applied here therefore promise a significantly upgraded performance of generation modules and, thereby, of natural language interaction in general.

Acknowledgements We would like to thank Simon Keizer, Helen Hastie and Xingkun Liu for collaboration on the real-user evaluations for information presentation.

The research leading to these results has received partial support from the European Community's Seventh Framework Programme (FP7) under grant agreement no. 216594 (CLASSIC project), from the EPSRC, project no. EP/G069840/1, and from the European Community's Seventh Framework Programme (FP7) under grant agreement no. 269427 (STAC project), under grant agreement no. 270019 (SPACEBOOK project), under grant agreement no. 270435 (JAMES project), and under grant agreement no. 287615 (PARLANCE project).

References

1. Akiba, T., Tanaka, H.: A Bayesian approach for User Modelling in Dialogue Systems. In: Proceedings of the 15th conference on Computational Linguistics - Volume 2, Kyoto (1994)
2. Angeli, G., Liang, P., Klein, D.: A simple domain-independent probabilistic approach to generation. In: Proceedings of EMNLP (2010)
3. Bell, A.: Language style as audience design. Lang. Soc. **13**(2), 145–204 (1984)
4. Boidin, C., Rieser, V., Plas, L.V., Lemon, O., Chevelu, J.: Predicting how it sounds: Re-ranking alternative inputs to TTS using latent variables. In: Proc. of Interspeech/ICSLP, Special Session on Machine Learning for Adaptivity in Spoken Dialogue Systems (2009)
5. Cawsey, A.: User Modelling in Interactive Explanations. User Modeling and User-Adapted Interaction 3(3), 221–247 (1993)
6. Chung, G.: Developing a flexible spoken dialog system using simulation. In: Proc. of the Annual Meeting of the Association for Computational Linguistics (ACL) (2004)
7. Clark, H.H., Murphy, G.L.: Audience design in meaning and reference. In: LeNy, J.F., Kintsch, W. (eds.), Language and comprehension. Amsterdam: North-Holland Publishing Company (1982)
8. Clarkson, P.R., Rosenfeld, R.: Statistical Language Modeling Using the CMU-Cambridge Toolkit. In: Proc. of ESCA Eurospeech (1997)
9. Cuayáhuitl, H., Renals, S., Lemon, O., Shimodaira, H.: Human-Computer Dialogue Simulation Using Hidden Markov Models. In: Proc. of the IEEE workshop on Automatic Speech Recognition and Understanding (ASRU), pp. 290–295, San Juan, Puerto Rico, November (2005)
10. Dale, R.: Cooking up referring expressions. In: Proc. ACL-1989 (1989)
11. Demberg, V., Moore, J.: Information presentation in spoken dialogue systems. In: Proc. of the Conference of the European Chapter of the Association for Computational Linguistics (EACL) (2006)
12. Dethlefs, N., Cuayáhuitl, H.: Hierarchical reinforcement learning and hidden markov models for task-oriented natural language generation. In: Proc. of 49th Annual Meeting of the Association for Computational Linguistics, 2011
13. Dethlefs, N., Cuayáhuitl, H., Viethen, J.: Optimising natural language generation decision making for situated dialogue. In: Proc. of SIGdial Workshop on Discourse and Dialogue (2011)
14. Dethlefs, N., Cuayahuitl, H.: Combining Hierarchical Reinforcement Learning and Bayesian Networks for Natural Language Generation in Situated Dialogue. In: ENLG (2011)

15. Eckert, W., Levin, E., Pieraccini, R.: User modeling for spoken dialogue system evaluation. In: Proc. of the IEEE workshop on Automatic Speech Recognition and Understanding (ASRU), pp. 80–87. Santa Barbara, CA , USA, December 1997
16. Gasic, M., Keizer, S., Mairesse, F., Schatzmann, J., Thomson, B., Young, S.: Training and Evaluation of the HIS POMDP Dialogue System in Noise. In: Proc. of SIGdial Workshop on Discourse and Dialogue, 2008
17. Golland, D., Liang, P., Klein, D.: A game-theoretic approach to generating spatial descriptions,. In: Proceedings of EMNLP (2010)
18. Henderson, J., Lemon, O.: Mixture Model POMDPs for Efficient Handling of Uncertainty in Dialogue Management. In: Proc. of the Annual Meeting of the Association for Computational Linguistics (2008)
19. Isard, A., Oberlander, J., Matheson, C.: Speaking the users languages. IEEE Intell. Syst. Mag. **18**, 40–45 (2003)
20. Issacs, E.A., Clark, H.H.: References in conversations between experts and novices. J. Exp. Psychol. Gen. **116**, 26–37 (1987)
21. Janarthanam, S., Lemon, O.: Learning to Adapt to Unknown Users: Referring Expression Generation in Spoken Dialogue Systems. In: Proc. ACL'10 (2010)
22. Janarthanam, S., Lemon, O.: Learning to Adapt to Unknown Users: Referring Expression Generation in Spoken Dialogue Systems. In: Proceedings of the 48th Annual Meeting of the Association for Computational Linguistics (ACL), pp. 69–78. Uppsala, Sweden, July 2010
23. Janarthanam, S., Lemon, O.: A Wizard-of-Oz environment to study Referring Expression Generation in a Situated Spoken Dialogue Task. In: Proc. ENLG'09 (2009)
24. Janarthanam, S., Lemon, O.: The GRUVE Challenge: Generating Routes under Uncertainty in Virtual Environments. In: Proceedings of ENLG / Generation Challenges (2011)
25. Janarthanam, S., Lemon, O.: Adaptive referring expression generation in spoken dialogue systems: Evaluation with real users. In: Proceedings of SIGDIAL (2010)
26. Janarthanam, S. Lemon, O.: A Two-tier User Simulation Model for Reinforcement Learning of Adaptive Referring Expression Generation Policies. In: Proc. SigDial'09 (2009)
27. Janarthanam, S., Lemon, O.: A Two-tier User Simulation Model for Reinforcement Learning of Adaptive Referring Expression Generation Policies. In: Proc./ of the Annual SIGdial Meeting on Discourse and Dialogue (2009)
28. Janarthanam, S., Hastie, H., Lemon, O., Liu, X.: 'The day after the day after tomorrow? ' A machine learning approach to adaptive temporal expression generation: training and evaluation with real users. In: Proceedings of SIGDIAL, 2011
29. Jung, S., Lee, C., Kim, K., Jeong, M., Lee, G.G.: Data-driven user simulation for automated evaluation of spoken dialog systems. Comput. Speech. Lang. **23**, 479–509 (2009)
30. Koller, A., Petrick, R.: Experiences with planning for natural language generation. In: ICAPS (2008)
31. Koller, A., Stone, M.: Sentence generation as planning. In: Proceedings of ACL (2007)
32. Koller, A., Moore, J., di Eugenio, B., Lester, J., Stoia, L., Byron, D., Jon Oberlander, J., Striegnitz, K.: Shared task proposal: Instruction giving in virtual worlds. In: White, M., Dale, R. (eds.) Working group reports of the Workshop on Shared Tasks and Comparative Evaluation in Natural Language Generation (2007)
33. Lemon, O.: Learning what to say and how to say it: joint optimization of spoken dialogue management and Natural Language Generation, Computer Speech and Language, **25**(2), 210–221 (2011)
34. Lemon, O.: Adaptive natural language generation in dialogue using Reinforcement Learning. In: Proc. of the 12th SEMdial Workshop on on the Semantics and Pragmatics of Dialogues, London, UK, June (2008)
35. Lemon, O., Janarthanam, S., Rieser, V.: Generation under uncertainty. In: Proceedings of INLG / Generation Challenges (2010)
36. Mairesse, F. Gasic, M., Jurcicek, F., Keizer, S. Thomson, B., Yu, K., Young, S.: Phrase-based statistical language generation using graphical models and active learning. In: Proceedings of the 48th Annual Meeting of the Association for Computational Linguistics (ACL) (2010)

37. Moore Johanna, D., Foster, M.E., Lemon, O., White, M.: Generating tailored, comparative descriptions in spoken dialogue. In: Proceedings of the Seventeenth International Florida Artificial Intelligence Research Sociey Conference. AAAI Press (2004)
38. Nakatsu, C.: Learning contrastive connectives in sentence realization ranking. In: Proc. of SIGdial Workshop on Discourse and Dialogue (2008)
39. Paris, C.L.: Tailoring Object Descriptions to a User's Level of Expertise. Comput. Ling. **14**(3), 64–78 (1988)
40. Polifroni, J., Walker, M.: Intensional Summaries as Cooperative Responses in Dialogue Automation and Evaluation. In: Proceedings of the Annual Meeting of the Association for Computational Linguistics (ACL) (2008)
41. Polifroni, J., Walker, M.: Learning database content for spoken dialogue system design. In: Proc. of the IEEE/ACL workshop on Spoken Language Technology (SLT) (2006)
42. Reiter, E., Sripada, S., Hunter, J., Davy, I.: Choosing words in computer-generated weather forecasts. Artif. Intell. **167**, 137–169 (2005)
43. Rieser, V., Lemon, O., Liu, X.: Optimising Information Presentation for Spoken Dialogue Systems. In: Proceedings of the Annual Meeting of the Association for Computational Linguistics (ACL), pp. 1009–1018. Uppsala, Sweden, July 2010
44. Rieser, V., Lemon, O.: Natural Language Generation as Planning Under Uncertainty for Spoken Dialogue Systems. In: Proc. of the Conference of European Chapter of the Association for Computational Linguistics (EACL), 2009
45. Rieser, V., Keizer, S., Lemon, O., Liu, X.: Adaptive Information Presentation for Spoken Dialogue Systems: Evaluation with human subjects. In: Proceedings of ENLG (2011)
46. Schlangen, D., Skantze, G.: A General, Abstract Model of Incremental Dialogue Processing. Dialog. Discourse **2**(1), 710–718 (2011)
47. Shapiro, D., Langley, P.: Separating skills from preference: Using learning to program by reward. In: *Proc. ICML-02* (2002)
48. Shapiro, D., Langley, P.: Separating skills from preference: Using learning to program by reward. In: Proc. of the 19th International Conference on Machine Learning (ICML), pp. 570–577, Sydney, Australia, July (2002)
49. Skantze, G., Hjalmarsson, A.: Towards Incremental Speech Generation in Dialogue Systems. In: Proceedings of the 11th Annual SigDial Meeting on Discourse and Dialogue, Tokyo, Japan (2010)
50. Stent, A., Prasad, R., Walker, M.: Trainable sentence planning for complex information presentation in spoken dialog systems. In: Association for Computational Linguistics (2004)
51. Sutton, R., Barto, A.: Reinforcement Learning. MIT Press, Cambridge (1998)
52. van Deemter, K.: Utility and Language Generation: The case of Vagueness. Journal of Philosophical Logic 607–632 (2009)
53. van Deemter, K.: What game theory can do for NLG: the case of vague language. In: 12th European Workshop on Natural Language Generation (ENLG) (2009)
54. Walker, M., Passonneau, R., Boland, J.: Quantitative and qualitative evaluation of DARPA Communicator spoken dialogue systems. In: Proc. of the Annual Meeting of the Association for Computational Linguistics (ACL) (2001)
55. Walker, M., Stent, A., Mairesse, F., Prasad, R.: Individual and domain adaptation in sentence planning for dialogue. J. Artif. Intell. Res. (JAIR) **30**, 413–456 (2007)
56. Walker, M.A., Kamm, C.A., Litman, D.J.: Towards Developing General Models of Usability with PARADISE. Nat. Lang. Eng., **6**(3), 363–377 (2000)
57. White, M., Rajkumar, R., Martin, S.: Towards broad coverage surface realization with ccg. In: In Proceedings of the Workshop on Using Corpora for NLG: Language Generation and Machine Translation (UCNLG+MT (2007)
58. Whittaker, S., Walker, M., Moore, J., Whittaker, S., Walker, M., Moore, J.: Fish or fowl: A Wizard of Oz evaluation of dialogue strategies in the restaurant domain. In: Proc. of the International Conference on Language Resources and Evaluation (LREC) (2002)

59. Winterboer, A., Hu, J., Moore, J.D., Nass, C.: The influence of user tailoring and cognitive load on user performance in spoken dialogue systems. In: Proc. of the 10th International Conference of Spoken Language Processing (Interspeech/ICSLP) (2007)
60. The Parlance Project, www.parlance-project.eu, (2012)
61. Young, S.J., Schatzmann, J., Weilhammer, K., Ye, H.: The Hidden Information State Approach to Dialog Management. In: ICASSP 2007 (2007)
62. Young, S., Gašić, M., Keizer, S., Mairesse, F., Thomson, B., Yu, K.: The Hidden Information State model: a practical framework for POMDP based spoken dialogue management. Comput. Speech. Lang. **24**(2), 150–174 (2009)

Chapter 7
Metrics and Evaluation of Spoken Dialogue Systems

Helen Hastie

7.1 Introduction

The ultimate goal of an evaluation framework is to determine a dialogue system's *performance*, which can be defined as "the ability of a system to provide the function it has been designed for" [32]. Also important, particularly for industrial systems, is *dialogue quality* or *usability*. To measure usability, one can use subjective measures such as User Satisfaction or likelihood of future use. These subjective metrics are difficult to measure and are dependent on the context and the individual user, whose goal and values may differ from other users. This chapter will survey evaluation frameworks and discuss their advantages and disadvantages. We will examine metrics for evaluating system performance and dialogue quality. We will also discuss evaluation techniques that can be used to automatically detect problems in the dialogue, thus filtering out good dialogues and leaving poor dialogues for further evaluation and investigation [62].

As commercially deployed spoken dialogue systems (SDS) become more widespread (see Chap. 8), it is important to take into consideration business needs during evaluation. Table 7.1 lists the key stakeholders of deployed SDS, namely: the developer or researcher, the business operator and, of course, the end-user. The table lists key aspects of evaluation that each stakeholder is concerned with. The developer needs to make sure the system performs as it should do with a high degree of usability and is interested in how performance affects user-perceived quality. The business operator also needs to make sure the system functions as it should do but is also concerned with how the persona of the business is reflected in the system and in retaining customers through highly satisfactory systems and

H. Hastie (✉)
Heriot-Watt University, Edinburgh, EH14 4AS, UK
e-mail: h.hastie@hw.ac.uk

O. Lemon and O. Pietquin (eds.), *Data-Driven Methods for Adaptive Spoken Dialogue Systems*, DOI 10.1007/978-1-4614-4803-7_7,

Table 7.1 Stakeholders and aspects of evaluation important to them

Stakeholder	Key aspects of evaluation
Developer	Performance of system and components how performance affects perceived quality
Business operator	Performance, dialogue quality/usability, acceptability, future use/repeat calls, dialogue efficiency, system persona/branding, cost saving, ROI
End-user	Dialogue quality/usability, performance in terms of task completion, engagement/dialogue efficiency (depending on task)

repeat calls. As time is money, business operators are also typically interested in time on task and cost savings. Finally, the end-user may be calling for a specific reason or to perform a specific task (e.g. getting a bus time, making a bank transfer) and therefore is interested in task completion and having an efficient, effective and streamlined conversation. If the task is more of an exploratory one, such as an interactive museum guide, then rather than dialogue efficiency, user engagement is possibly a more appropriate measure.

There is a strong relation between data-driven statistical optimisation of dialogue systems and the field of SDS evaluation. Indeed, one major advantage of data-driven approaches, such as reinforcement learning, is that strategies can be optimised and evaluated using the same objective/reward function [43, 60]. One needs to define certain metrics to be optimised in the reward function and evaluation frameworks can help identify these metrics. Learning which parameters to include in the reward function is preferable to hand-coding or best-guessing [35]. In addition, prior to costly testing with humans and to optimise strategies, evaluating systems using user simulations can provide a cost-effective and efficient means of system evaluation (see Chap. 4). The reward function and optimisation of statistical systems using simulated dialogues are intrinsically linked and rely on the ability to effectively evaluate the quality of a dialogue or interaction. This reward function can also be influenced by business constraints for industrial systems, such as in Speech Cycle's Contender system [50] where call automation and time on task are used in the reward function (see Chap. 8 for a full description of Contender).

Typically, evaluation has been in the form of task-based exercises where participants are paid a fixed fee and asked to perform a task in the laboratory and then fill out a usability questionnaire. This is not only highly inefficient, it also creates unrealistic scenarios and user goals. The field is moving more towards evaluations "in the wild" such as the spoken dialogue challenge (SDC) Let's Go shared task [4]. With these types of evaluations, collecting subjective metrics such as dialogue quality is very challenging as real, busy users are likely to hang up once they have the information they need, rather than staying on the line to fill out a questionnaire. We will discuss throughout this chapter how evaluation frameworks can be applied to real tasks "in the wild".

7.2 Evaluation Frameworks

In the 1980s and 1990s, systems were evaluated in terms of individual metrics such as turn correction ratio, concept accuracy, implicit recovery and transaction success [8, 19]. Rather than looking at individual metrics, one of the first complete evaluation frameworks to emerge was PARADISE (PARAdigm for DIaLogue System Evaluation) [57]. PARADISE aims to examine the delicate balance and trade-off of a variety of metrics with the aim of optimising a response variable such as User Satisfaction. In this section, we describe PARADISE and other frameworks including the more recently proposed tripartite framework by Möller and Ward [33].

7.2.1 PARADISE

The PARADISE [57] framework has been widely adopted in SDS evaluations [5, 27, 40]. It was initially used in the 2000 and 2001 DARPA Communicator shared task in the domain of travel booking. The PARADISE evaluation framework uses methods from decision theory [26] to combine a disparate set of performance measures (i.e. User Satisfaction, Task Success and Dialogue Cost) into a single performance evaluation function [57]. The use of decision theory requires a specification of both the objectives of the decision problem and a set of measures (known as attributes in decision theory) for operationalising the objectives.

The PARADISE model posits that performance can be correlated with a meaningful external criterion such as usability and thus that the overall goal of a spoken dialogue agent is to maximise an objective related to usability. The model further posits that two types of factors are potentially relevant contributors to User Satisfaction (namely Task Success and Dialogue Costs) and that two types of factors are potential relevant contributors to costs (namely efficiency measures and dialogue quality measures). PARADISE uses multiple linear regression to quantify the relative contribution of these predictors reflected in their coefficients.

In the early 1990s, User Satisfaction ratings were frequently used in the literature [25,48] as an external indicator of the usability of a dialogue agent. User Satisfaction is typically calculated by summing the answers to the following five questions answered on a seven point Likert scale: TTS performance, ASR performance, task ease, expected behaviour and future use. Further discussion of User Satisfaction and other subjective metrics is given in Sect. 7.3.2.

Several uses have been made of the models derived by applying PARADISE. Firstly, PARADISE has been used for automatic prediction of problematic dialogues [52, 62]. Secondly, the weights from the linear model have been used to define reward functions for data-driven approaches to dialogue management [18,43,59].

Table 7.2 PARADISE-derived linear model for the 2000 communicator evaluation [55]

Factor	Coefficient
Task completion	0.43
Task duration	−1.5
Sentence accuracy	0.21
System turn duration	0.14

7.2.1.1 PARADISE Case Study

The PARADISE evaluation framework was applied to the DARPA Communicator challenge in two stages performed in 2000 and 2001 involving nine and eight participating sites, respectively [53–56]. This shared task was one of the first to attempt to use subjects with realistic goals, from a target population of frequent travellers. In the 2000 experiment, task scenarios consisted of seven fixed and two open scenarios. The fixed scenarios were designed to vary task complexity where task complexity was defined as the number of goals the user had (e.g. hotel, car hire). The open scenarios were defined by the user. As discussed by [56], having the users define their own tasks did not give the expected results as the users tended to change their task if faced with speech recognition errors.

In the 2001 data collection, the subjects had to make travel arrangements with varying complexity (round vs. multi-leg, fixed airline, car hire, hotel). This experiment ran for 6 months where the system was continuously accessible via a toll-free number. As well as fixed scenarios, the participants had to make *real* arrangements for their own travel. This was very similar to evaluations "in the wild" that the community is now gravitating towards (e.g. the SDC Let's Go challenge [4]).

Example metrics collected in the Communicator evaluation are listed below and taken from [55]. A more detailed discussion of metrics is given in Sect. 7.3:

- **Dialogue Efficiency**: Total elapsed time, task duration, num. system/user turns, num. turns on task, system/user turn duration, System words per turn
- **Dialogue Quality**: Word accuracy, sentence accuracy, mean response latency, response latency variance, num. of Overlaps
- **Task Success**: Perceived task completion, exact scenario completion, any scenario completion
- **User Satisfaction**: Sum of TTS performance, task ease, user expertise, expected behaviour and future use

The evaluation of the 2000 Communicator systems resulted in a learned model with the coefficients given in Table 7.2 taken from [55]. This model accounts for 35 % of the variance. The learned model for the 2001 Communicator weights is given in Table 7.3 which accounts for 27 % of the variance taken from [56].

Although the sets of metrics were calculated differently, in both these learned models there are four common themes in the metrics that contribute to User Satisfaction. Firstly, in both evaluations, task completion is important. Secondly,

Table 7.3 PARADISE-derived linear model for the 2001 communicator evaluation [56]

Factor	Coefficient
Task duration	−0.44
System words per turn	0.41
Task completion	0.32
Sentence error rate	−0.17
Number of overlaps	−0.15
User words per turn	0.04

metrics that capture length are important for both evaluations. Thirdly, a measure of accuracy is included in both models (i.e. sentence error rate (SER) or sentence accuracy). Finally, in both evaluations, longer system turn duration is preferred, the hypothesis being that instructions and itinerary presentations tend to be longer indicating a certain level of success. The difference of the models' fits can be accounted for by the greater variability in the 2001 data.

A further evaluation reported in [54] resulted in an increased fit of the 2000 model from 37 % to 42 % by including system dialogue acts using the DATE dialogue act tagging schema [58]. This tagging scheme is described in further detail in Sect. 7.3. Here, again the most predictive metrics include not only task completion, task duration, system turn duration and word accuracy but also certain dialogue acts such as giving a price or making a booking. One interpretation of the metrics given by [54] is that these dialogue acts are acting as landmarks in the dialogue indicating completion of a particular set of subtasks. For example, "TripC" is a metric that counts the number of open-ended questions about the user's travel plans. These dialogue acts typically occur at the beginning of the dialogue but also after one itinerary has been planned. Another metric is "ReqDate" that counts utterances such as *Could you tell me what date you want to travel* which typically occur after the origin and destination have been established. These dialogue act metrics have also been used for automatic prediction of problematic dialogues which is described in further detail in Sect. 7.4.

Task completion or Task Success has a positive influence on User Satisfaction. However, this relationship is not always linear. For example, [62] discuss the relationship between Task Success and User Satisfaction. It can be the case that User Satisfaction is high, but Task Success is low. In this case, the user may have enjoyed the dialogue but did not end up completing the task. Conversely, there are cases where Task Success is high, but User Satisfaction is low; in these dialogues, the user had to frequently repeat themselves, and the dialogue may have taken many turns, but eventually the user completes the task. Though successful, these dialogues obviously have low usability.

Some metrics are consistently found to be contributors to User Satisfaction across systems. For example, a negative effect of task duration was also found in the evaluation of two systems reported in [61]: (1) ANNIE: a system that provides facilities for voice dialling, employee directory look-up and voice and email access and (2) ELVIS-SI: a system-initiative version of the ELVIS system for accessing email. Walker et al. [61] describe the relationship between elapsed time (ET), i.e. the

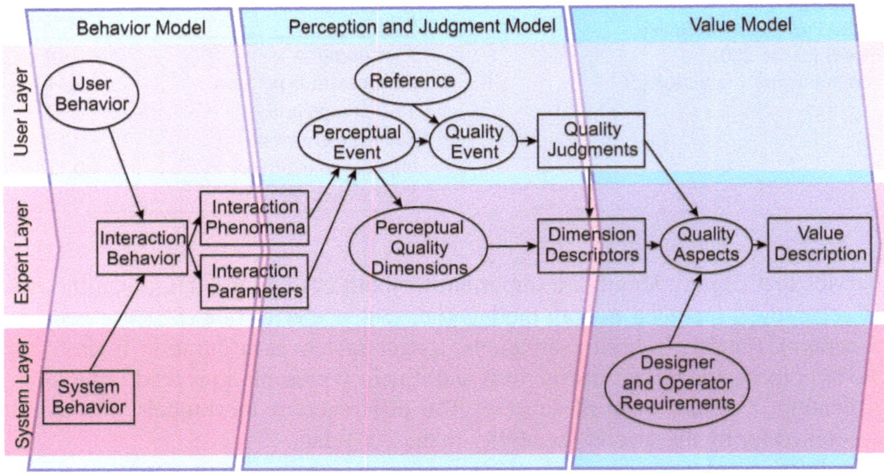

Fig. 7.1 Möller and Ward tripartite evaluation framework taken from [33]

length of the interaction, and other metrics including User Satisfaction. They found that ET cannot be used to replace dialogue quality metrics and that the importance of ET increases for dialogues where the user was successful in completing the task. However, they showed that the importance of the evaluation metric ET does not increase as recognition performance increases as one might expect. They also showed that alternate measures of User Satisfaction such as task ease, which are more biased towards finding a relation with lapsed time, are also highly correlated with qualitative measures and Task Success. This illustrates the need for all types of metrics and confirms that the evaluation of SDSs is not easily simplified nor the number of metrics easily reduced.

7.2.2 Möller and Ward

Möller and Ward [33] posit that quality judgement involves a perception process where the user "compares the perceptual event with respect to some (typically implicit) reference". Both the judgement and perception processes are events which happen in a particular personal, spatial, temporal and functional context. They propose a tripartite framework with one part modelling the behaviour of the user and system during the interaction, the second part the perception and judgement processes taking place inside the user and the third part models what matters to system designers and service providers. This framework is illustrated in Fig. 7.1 from [33].

Firstly, the *behaviour model* translates the characteristics of the system and the user into predicted interaction behaviour. One way to do this is to collect "interaction parameters" which quantify the behaviour of the interaction participants in a similar way to PARADISE's efficiency measures (see Sect. 7.3 for further details). According to the model described in [33], interaction phenomena fall in this category, such as system errors, user errors, points of confusion, and dead time. Möller and Ward [33] discuss running user simulations to obtain values for the behaviour model thus avoiding using real participants. Evaluations with user simulations are discussed in more detail in the following section.

The second model of the framework is the *perception and judgement model* which captures the quality perceived by the user which is analogous to the User Satisfaction metric used in the PARADISE model. Specifically, the user is asked to describe the *perceptual event* or the *quality event* either qualitatively through open questions or quantitatively on rating scales. A detailed discussion on how to capture dialogue quality is described in Sect. 7.3.2.

Möller and Ward [33] state that the stakeholders in Table 7.1 may focus on different quality dimensions and weigh them differently based on the current task, context of use, price, etc. Their *value model* derives a *value description* which takes weighting of aspects into account. An example they give in [33] is for a system to teach girls to "talk to Barbie" where speech recognition performance or completion of task is of less importance than voice quality, responsiveness and unpredictability. This system would have a very different value description to defence applications, for example. The value function is also important for the business operator and would vary from system to system and business to business. Predicting this value function is very challenging as it depends on a company's business model and customer base and may even be proprietary.

In summary, this model not only takes into account what is important for the developer but also discusses a value model of perceived quality. This would lead to an understanding of what is important for the target user group which could be inferred from previous history, user profiles, etc., and lead to systems that adapt to individual users, for example, adapting to novices vs. experts [23]. Möller and Ward [33] claim that this evaluation framework should work for systems of any type and enable more accurate prediction of the value of design innovations as well as incremental system improvements.

7.2.3 Dialogue Simulations

With the cost and effort to perform full evaluations and with the advent of statistical SDSs, many groups are opting for testing using simulated dialogues [10] (see Chap. 4 for an in-depth discussion on user simulations). This facilitates optimisation and irons out any system errors prior to testing with real users as simulated users can explore large possibilities of user reactions [41, 45]. When evaluating systems or strategies with user simulations, typically the cumulative reward is used as the metric which determines the best performance. Evaluation of data-driven strategies both for dialogue strategies [63] and individual components such as natural language generation (NLG) [24] is becoming commonplace.

User simulation environments vary according to their level of abstraction which can either be at the signal, word or intention level. Intention-level simulation is often used for evaluating dialogue and NLG strategies as it avoids the complexity and variability of natural language [10, 28, 37]. Signal-level dialogue simulation is rarer [30], but does allow for testing of the recognition and understanding modules. User simulations can be further classified as either deterministic, i.e. rules based on heuristics observed in the data [30], or statistical, i.e. trained on observed data. These data can be human–human data or human–computer data [7, 14], Wizard of Oz data (see Sect. 7.2.5 for discussion) [23, 44] or even data from another SDS [2].

Deterministic models have been used to evaluate which dialogue strategies work well for different types of user response patterns [29]. In general, probabilistic models are more realistic for modelling user behaviour and closer to an evaluation with real users. Memo [31] models the interaction behaviour by a so-called "mental model" to emulate user errors based on a general domain model, user studies and a system task model. Using the Memo toolkit, [12] performed a comparison between a simulated corpus and a human–computer corpus with both interaction sets being used to compare two user groups (young and old) and two system versions. They found that using the user simulation data, they could rank the four experimental conditions for different interaction parameters. That is to say, they were able to predict which version of the system was rated better by which group of users. Further discussion of automatic evaluation is discussed in Sect. 7.4.

The study reported in [1] is based on highly subjective questions asked to the human judges observing dialogues between a student and a tutor. It subsequently uses the scores provided by human judges to train different metrics with supervised learning methods (stepwise multiple linear regression and ranking models). The study concludes that the latter method is able to mimic correctly human judgements and could be used to evaluate new simulated dialogues. This method is closely related to PARADISE discussed above.

Another evaluation platform that uses simulated dialogues is SpeechEval [46], the goal of which is to use a realistic user model for system evaluation. The quality of the simulated dialogues that are used in the evaluation is key to the accuracy of the evaluation of the dialogue system being tested. For example, if the user simulation is supposed to model novices vs experts [23] or young vs. old [12], the model must be a fair representation of the user group otherwise the findings inferred from the evaluation may be false. Evaluating the user simulations themselves is a topic in itself for discussion and outside the scope of this chapter (see [38] for a survey of user simulation evaluation metrics and Chap. 4).

7.2.4 SERVQUAL

Hartikainen et al. [17] describe a purely subjective framework for evaluation, adapting the SERVQUAL (SERVice QUALity) evaluation method which is a service quality evaluation method developed by marketing academics. They claim that service quality can be divided in dimensions and that this method can measure the difference of expectations and perceptions of quality. The service quality dimensions are *reliability* (ability to perform the promised service dependably and accurately), *responsiveness* (willingness to help customers and provide prompt service), *assurance* (employees' knowledge and courtesy and their ability to inspire trust and confidence), *tangibles* (appearance of physical facilities, equipment, personnel and other materials) and *empathy* (caring, individualised attention given to customers). Pre- and post-test questionnaires are delivered to participants with the pre-test extracting the participants' expectations and the post-test their quality judgements. This goes someway to capturing Möller and Ward's value model. Hartikainen et al. [17] tested their evaluation framework on an email system, and they discuss different user group value priorities such as the importance of system performance of males vs. females and the zone of tolerance of system reliability. Being a purely subjective evaluation methodology, this framework provides an alternative to User Satisfaction and provides someway to defining a value model; however, to be a complete evaluation framework, it would need to be combined with objective measures as is the case with the PARADISE and Möller and Ward's frameworks.

7.2.5 "Wizard of Oz" Evaluation

A framework for evaluation which is particularly useful early on in development is Wizard of Oz or WOZ experimentation. This is also often used to maintain a gold standard to which subsequent system versions can be compared [34]. In a WOZ set-up, there is a human wizard who usually does not hear the user but receives their output as text and responds as text (or a dialogue act that is synthesised as text) which the user hears. Therefore, the user interacts as they would with a working system. WOZ is a powerful mechanism for testing certain strategies, such as different referring expressions [23] and comparing resulting evaluation metrics (e.g. Task Success, time on task). The major drawback of WOZ experiments is that the WOZ environments are costly to build and running experiments and training the wizards are time consuming. In addition, there may be discrepancies between the different styles of wizards. Despite these drawbacks, having a gold standard that allows for system optimization and that can also be used to bootstrap reinforcement-based systems is very valuable [42].

7.2.6 Discussion of Evaluation Frameworks

Möller and Ward [33] argue that the PARADISE model does not reliably give valid predictions of individual user judgments, typically covering only around 40–50 % of the variance in the data it is trained on. However, PARADISE has been shown to be able to successfully predict dialogue quality automatically [62] and also as a useful mechanism for determining reward functions [43,60]. Evaluation frameworks in general may be overambitious, attempting to link dialogue quality to a single response variable such as User Satisfaction, and perhaps a response variable based on perceptual quality judgments would be more appropriate. Alternatives to User Satisfaction are discussed further in Sect. 7.3.2. Furthermore, linear combination of factors may not be a good fit for all features. For example, a very low number of system turns may be an indication of problematic dialogues as is a very large number of system turns.

The difference between PARADISE and Möller and Ward's model is that the latter includes a value model, essentially a user model. The Möller and Ward model not only includes efficiency measures and User Satisfaction but also provides feature weights based on what is important for the stakeholders. However, Möller and Ward's model has yet to be developed and tested in full.

7.3 Metrics

In this section, we describe the wide range of performance metrics used in the literature to capture system *performance* and *dialogue quality*. There are standard metrics to gauge performance of individual components that are easily computable and interpretable, such as word error rate (WER) for the automatic speech recognition (ASR) module or concept accuracy (CA)/concept error rate (CER) for the natural language understanding (NLU) module. Other components are not so easy to evaluate, such as the dialogue manager and NLG. Individual components contribute to system performance as a whole, but evaluating them individually is not always necessary when determining system performance. Here, we do not attempt to list all metrics for each module of a SDS, rather we discuss metrics that capture overall system performance and dialogue quality.

PARADISE [57] categorises metrics in terms of Task Success and Dialogue Costs, with costs being divided into dialogue efficiency measures (e.g. task duration, system words per turn) and dialogue quality measure (e.g. WER, sentence accuracy). In this chapter, we discuss metrics in terms of being subjective and objective. Figure 7.2 shows how subjective measures (i.e. the user's point of view) include dialogue quality such as User Satisfaction and also subjective Task Success (i.e. asking the user if they completed the task or not). Objective measures are in terms of reducing Dialogue Costs as in the PARADISE model, but here we also include objective Task Success (i.e. did they complete their task successfully?). Figure 7.2 is not intended to be an evaluation framework but rather a means for classifying different metrics into the different types for ease of discussion.

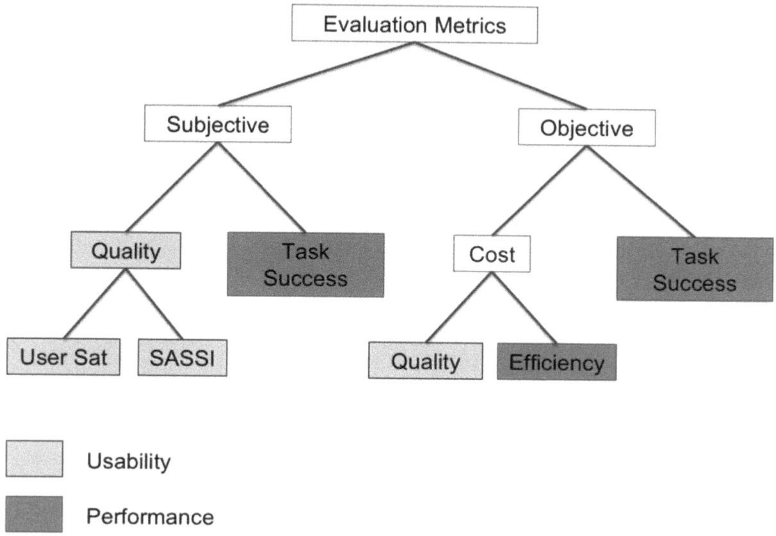

Fig. 7.2 Figure representing the partition of evaluation metrics where *dark grey* boxes represent system *performance* and *light grey dialogue quality* or *usability*

7.3.1 Objective Metrics

As seen in Fig. 7.2 for objective measures, we follow the PARADISE framework categorising objective metrics in terms of cost (dialogue efficiency, dialogue quality) and objective Task Success. Examples of these metrics are given below:

- **Dialogue Efficiency**: Total elapsed time, time on task, system turns, user turns, total turns, time per turn for each system module, average words per user turn, average words per system turn.
- **Dialogue Quality**: WER, number of ASR rejects, number of overlaps in system/user turns.
- **Task Success**: objective measures from logs.

The International Telecommunication Union-Telecommunication Standardization Sector (ITU-T) proposes a set of objective interaction parameters targeted for the evaluation of telephone-based SDSs [21]. Here the parameters are categorised into the following five types listed below with example metrics (the reader is referred to [21] for a full list of metrics):

1. Dialogue and communication-related parameters (e.g. dialogue duration, system/user turn duration, system response delay, query density, concept efficiency).
2. Meta-communication-related parameters (e.g. #help-requests, #time-out, #system error).
3. Cooperativity-related parameters (e.g. contextual appropriateness).

4. Task-related parameters (e.g. Task Success, kappa coefficient).
5. Speech-input-related parameters (e.g. WER, SER, concept accuracy (CA)).

The dialogue and communication-related parameters refer to the overall dialogue and to the communication of information and give a rough indication of how the interaction takes place. Parameters include duration-related parameters and word- and turn-related parameters. This category also includes query density (how efficiently a user can provide new information) and concept efficiency (how efficiently the system can absorb information). This category is similar to the dialogue efficiency category in the PARADISE framework.

The second category is meta-communication-related parameters, such as corrections, requests for help, barge-ins and clarifications and sub-dialogues to recover from misunderstandings. This category is a mixture of system dialogue act counts and dialogue quality measures. The third category attempts to capture the co-operativity of the system where they suggest classifying dialogues by hand for appropriateness (not violating Grice's maxims [16]), inappropriateness (violating one or more maxims), incomprehensibility (meaning of utterance cannot be discerned) or total failure (no system response).

The fourth category is for task-related parameters such as Task Success and kappa coefficient. Task Success is a common parameter among all evaluation frameworks. Typically, users are given a task to perform and are asked for their opinion on whether the task was completed (*subjective* Task Success). In addition, *objective* Task Success can be gleaned from the logs or by hand-labelling the dialogues. With evaluations "in the wild" where users are able to determine their own goal, automatic objective Task Success extraction is a difficult issue especially if the user modifies their goal mid-dialogue. For example, the user starts looking for cheap Italian restaurants in the centre of town but changes their mind slightly and widens their search criteria. If they find a restaurant and are pleased with the result then that could be classified as Task Success even if their initial criteria changed during the dialogue.

The final category of parameters from ITU-T include Speech-input-related parameters which come under dialogue quality in the PARADISE framework. These parameters include WER, SER and number of errors per sentence among others. Much emphasis has been placed on metrics such as WER in the past; however, as systems are being developed that are more able to adapt to noisy environments and as a consequence to poor speech recognition quality, metrics such as CER are perhaps more reflective of actual dialogue quality. Young et al. [63] evaluate different dialogue strategies with user simulations by plotting average success and reward rates as a function of error rate. These graphs can illustrate the robustness of the systems when faced with decreasing semantic accuracy.

Finally, [51] list a set of metrics for measuring performance of a commercial system in terms of completion rate, average holding time, hang-up and opt-out rates, retry rates and caller experience. These can be classified in terms of Fig. 7.2 as objective measures but with one subjective measure of caller experience measured on a scale of 1–5.

As well as the metrics outlined above, unigram counts of system dialogue acts can also be included in the evaluation as an objective measure of quality. For shared task evaluations such as the communicator tasks, dialogue act schemas across systems can vary immensely. To compensate for this, dialogue act parsers such as that described in [58] can be used to label cross-site logs using the same dialogue act schema. Walker et al. [54] show that including system dialogue act counts increases the response variable coverage by 5 %. The dialogue act schema adopted by [54] is known as DATE [58]. Subsequent dialogue act schemas that have been successfully used in dialogue systems and could be used for evaluation in a similar way include the CUED-based dialogue act schema used on the TALK and CLASSiC projects [39].

7.3.2 Subjective Metrics

Typically, SDS are evaluated in terms of the performance of the system and user-perceived quality judgements [20, 33, 57]. The definition of dialogue quality, as accepted by ITU-T [22], is defined as "the result of judgement of the perceived composition of an entity with respect to its desired composition". As observed by [33], it is this quality that involves both a perception process and a judgement process whereby the person compares the perceptual event with a (typically implicit) reference event. Möller and Ward [33] and others argue that this comparison with a reference is associated with a user-specific value of the perceptual event which takes place in a particular context of use. Being a highly subjective measure, users have different preconceived ideas of what a system should be able to do and have different expectations. It is the gathering of these perceived quality judgements that is very challenging especially for evaluations "in the wild" since the user is not motivated through compensation to give feedback as with lab-based experiments and will likely hang up when they have the information they require from the SDS. Similarly, it is very difficult to rank systems by asking the user for preferences of more than one system.

As discussed in Sect. 7.2.1, the PARADISE framework requires a questionnaire of five questions on a seven-point Likert scale gathering the following criteria: TTS performance, ASR performance, task ease, expected behaviour and future use. This User Satisfaction metric is taken as the overall objective to be maximised by a system and that Task Success and various interaction costs can be used as predictors of User Satisfaction. Möller and Ward [33] posit that it is overambitious to directly relate a single metric to a measure of overall system quality. Rather it is better to limit the scope of the perception and judgement component to the prediction of values on a number of perceptual quality dimensions. Obtaining values on these dimensions, however, requires administrating lengthy questionnaires which may not always be possible. In addition, judgements can change depending on the order the systems are evaluated [9].

One such questionnaire is subjective assessment of speech system interfaces (SASSI) [20] which consists of 50 declarative statements. Factor analysis reveals six main factors including system response accuracy, likeability, cognitive demand, annoyance, habitability and speed [20]. Another questionnaire is ITU-T Rec P.851 (2003) [22] which has three separate (1) questionnaire distributed at the beginning of the evaluation for collecting information on the user's background, (2) questionnaire with questions related to individual interactions with the system under test and (3) questionnaires related to the user's overall impression of the system, to be answered after a number of interactions with the system using a 5-point Likert scale or on continuous rating scales. It has been shown that both SASSI and ITU-T questionnaires can cover a large number of different quality usability dimensions with a high validity and reliability; however, clearly it is unrealistic to administer three different types of questionnaires or indeed ask the real user/customer to answer 44 questions for deployed systems.

7.3.3 Discussion

Here, we have described a wide variety of metrics both subjective and objective. It is clear from the evaluation frameworks described in Sect. 7.2 that evaluation has to take into account both types of metrics and the weights of which can be learned to determine what exactly contributes negatively and positively to the user experience. Both types of metrics have their advantages and disadvantages. For example, User Satisfaction is highly subjective and can vary based on context and user, whereas objective quantifiable metrics may be difficult to interpret due to their non-linear nature.

7.4 Problematic Dialogue Detection

This book focuses on the optimisation of systems under uncertainty, for example in noisy environments. In these conditions, problems such as misrecognitions and misunderstandings can lead to user frustration and, on occasion, increased user speech intensity and as a result further increase in ASR uncertainty. Part of the problem is detecting when dialogues are starting this downwards spiral and then learning strategies that can mitigate these situations. For example, with commercial systems, if a problem is detected early enough during the conversation, then the customer can be transferred to a customer care representative or the dialogue system could modify its dialogue strategy in an attempt to repair the problem (see Chap. 8).

Problematic dialogue detection has been an active field since the late 1990s and can also be applied off-line for dialogue logs analysis, acting as a filter, filtering out the good dialogues and leaving the poor ones for further analysis [52, 62]. Most commercial SDSs adopt a continuous optimisation process. Through automatic quality prediction, problematic dialogues could be targeted for listening and transcribing rather than just picking dialogues randomly, saving time and money.

The linear regression function derived through PARADISE as discussed in Sect. 7.2.1 can be used to predict User Satisfaction with unseen dialogues [57]. An extension of this is using contributing metrics identified through PARADISE as input to supervised learning algorithms (e.g. CART, RIPPER) for the classification of dialogues in the domains of travel planning (DARPA Communicator) [62] and customer care centres [52]. These systems perform 26 % and 13.2 % over and above the majority class baseline, respectively. These classifiers use dialogue acts which act as landmarks in the dialogue. For example, if a dialogue act related to ground arrangements is observed, this indicates that the user has successfully completed the flight booking part of the dialogue. What impact does this work have in terms of transcription time and resources saved? Wright-Hastie et al. [62] discuss that, if one had a budget to transcribe 20 % of their dataset containing 100 dialogues, then by randomly extracting 20 dialogues, one would transcribe five problematic dialogues and 15 good dialogues. On the other hand, using their fully automatic problematic dialogue detector, one would have obtained 12 problematic dialogues and eight good dialogues. To look at it in another way, to extract 15 problematic dialogues out of 100, 55 % of the data would need transcribing. To obtain 15 problematic dialogues using their problematic dialogue predictor, only 26 % of the data would need transcribing.

Schmitt et al. [47] and [11] have looked at predicting User Satisfaction on a turn-by-turn basis. Engelbrecht et al. [11] describe models that can predict a distribution of ratings which can be expected for a group of users, and also they consider time relations between events, modelling user opinion as a variable over the course of a dialogue. To do this, they use hidden Markov models (HMMs) with reasonable success. Engelbrecht et al. [11] create an artificial environment where the interaction is paused at each turn so that the user can rate the previous turn on a 5-point scale of "bad", "poor", "fair, "good" and excellent". These data are then used to train the HMM to model User Satisfaction. Schmitt et al. [47] use third-party annotators to rate each turn of the dialogue and train a support vector machine (SVM) classifier to discriminate between five classes of "good" to "very poor", reaching a performance of 61.6 % unweighted average recall (also known as match rate per rating).

It is not just User Satisfaction that can be used for inferring problematic dialogues. Rieser and Lemon [43] use a multiple linear regression model for task ease rather than User Satisfaction and use this to predict task ease for unseen dialogues. Ideally, future work would combine overall problematic dialogue prediction with temporal features and possibly combine a notion of a user model to normalise for user variation.

7.5 Evaluation of Commercial Systems

There is a clear disjunction of evaluation goals of industrial deployed systems and research prototypes [35, 36]. For industry, usability, cost and return on investment (ROI) are the primary goals whereas naturalness of interaction and freedom of expression are the primary goals of research [35]. Typically the latter does not necessarily lead to the former. For example, a directed dialogue may have less freedom of expression; however, it may have higher usability and Task Success as the user has a good idea of what he or she should be saying at any one time. Domains where one frequently sees commercial voice user interfaces (VUIs) include structured tasks such as travel booking, banking and obtaining flight information that require a standard set of steps (e.g. for flight information: airline, flight number and destination). In fact, some VUIs are designed to follow a strict script that is also followed by their customer care service representatives. However, in some cases, unconstrained natural language is preferred, such as open How May I Help You (HMIHY) prompts [15] when there are too many options to list or the user may not have a good idea of which option they want.

Paek [35] defines the problem as one of commensurability for industrial and research systems. This chapter has discussed a number of frameworks and metrics that seek a measure of commensurability, i.e. two or more systems can be measured by the same units. This is particularly difficult for systems that vary in so many different parts and in so many different ways, for example, different recognizers, different language models and different prompts. Commensurability is also a problem in industry; [35] gives the example of a system that adheres to best practices achieving a 90 % Task Success and savings of $500 million. Due to the fact that systems consist of so many different components and subcomponents, how does one know that this system is the optimal configuration?

Industry standards (e.g. VoiceXML, SRGS, EMMA etc.) and best practices go someway to determining optimality. As the field has matured over the years, so have best practices been accumulated in books [3, 6] and through platform providers' documentation and online training [13]. Best practices tend to be a list of rules or heuristics such as "use short prompts", "don't use open-ended prompts". These best practices have to be used with caution as they can be based on hunches and intuition and also may result in systems "of the same breed" that are not able to differentiate themselves. It is important to note that best practices are not limited to industry as the research community also have their set of best practices and research experiments led by the academic community and are also often cited by industry (see Chap. 8 for further discussion on best practices).

Paek [35] calls for a pooling of data from both research and industry to conduct meta-analysis which he defines as "a statistical analysis of a large collection of results from individual studies for the purpose of integrating the findings". In addition, [36] call for a "synergistic convergence" of the two approaches so that results from research can be channelled more easily into industry.

7.6 Discussion and Conclusion

We have discussed how evaluation is important, not only for system development but also for determining reward functions for statistical SDS and dialogue strategy optimisation. If one has no clear idea of what constitutes a good interaction then deciding on a goal for optimisation is difficult.

We have discussed aspects of evaluation that are important for all the stakeholders: the developer/researcher who needs to optimise performance with respect to perceived dialogue quality, the business operator who must get ROI and have a system that is representative of the company's image and values and finally, of course, the end-user who needs to not only accomplish a task but also, in a highly critical consumer society, have an enjoyable experience.

We have discussed a number evaluation frameworks such as PARADISE and Möller and Ward's framework. There frameworks attempt to capture overall system quality as a weighted combination of various metrics. Unfortunately, the research and industrial communities have yet to adopt a common evaluation platform. This may be due to the difficulty and cost of obtaining subjective measures such as User Satisfaction and subjective Task Success. As more and more systems are being evaluated "in the wild" rather than in the lab with fixed tasks, obtaining subjective measures of usability and dialogue quality is very challenging. Even with task-based evaluations, subjective measures are subject to variation based on the different users' context, attitude and values.

Finally, it is clear that the industry and research communities need to leverage each others' work going forward. Industry standards go some way to making this possible, but it is really the sharing of open data (particularly transcribed data) that is key to the field progressing. Many projects are aware of this and are starting to share their data (e.g. the CLASSiC project[1]). However, it is clear that a large effort would be needed to perform a statistical analysis across many data sets from industry and academia. These issues are currently being explored in the EC project PARLANCE[2], which aims to develop an "in the wild" evaluation framework for statistical incremental systems.

Acknowledgements I would like to acknowledge Olivier Pietquin and Oliver Lemon for their guidance in writing this chapter. The research leading to this work has received funding from the EC's FP7 programmes: (FP7/2007–13) under grant agreement no. 216594 (CLASSiC); (FP7/2011–14) under grant agreement no. 248765 (Help4Mood); (FP7/2011–14) under grant agreement no. 287615 (PARLANCE).

[1] http://www.classic-project.org

[2] http://www.parlance-project.eu

References

1. Ai, H., Litman, D.: Assessing dialog system user simulation evaluation measures using human judges. In: Proceedings of ACL, Columbus, Ohio (USA), pp. 622–629 (2008)
2. Araki, M., Doshita, S.: Automatic evaluation environment for spoken dialogue systems. In: ECAI Workshop on Dialogue Processing in Spoken Language Systems'96, pp. 183–194 (1996)
3. Balentine, B., Morgan, D.P.: *How to Build a Speech Recognition Application: A Style Guide for Telephony Dialogues*. Enterprise Integration Group (2002)
4. Black, A.W., Burger, S., Conkie, A., Hastie, H., Keizer, S., Lemon, O., Merigaud, N., Gabriel Parent, G., Schubiner, G., Thomson, B., Williams, J.D., Yu, K., Young, S., Eskenazi, M.: Spoken dialog challenge 2010: Comparison of live and control test results. In: Proceedings of the SIGdial (2011)
5. Bonneau-Maynard, H., Devillers, L., Rosse, S.: Predictive performance of dialog systems. In: Proceedings of the Language Resources and Evaluation Conference (LREC) (2000)
6. Cohen, M.H., Giangola, J.P., Balogh, J.: *Voice User Interface Design*. Addison Wesley Longman Publishing Co., Inc., Redwood City, CA, USA (2004)
7. Cuayhuitl, H., Renals, S., Lemon, O., Shimodaira, H.: Human-computer dialogue simulation using hidden markov models. In: Proceedings of ASRU, pp. 290–295 (2005)
8. Danieli, M., Gerbino, E., Metrics for evaluating dialogue strategies in a spoken language system. *CoRR* (1996)
9. Devillers, L., Bonneau-maynard, H.: Evaluation of dialog strategies for a tourist information retrieval system. In: Proceedings of ICSLP, pp. 1187–1190 (1998)
10. Eckert, W., Levin, E., Pieraccini, R.: User modelling for spoken dialogue system evaluation. In: Proceedings of ASRU, pp. 80–87 (1997)
11. Engelbrecht, K.P.: Gödde, F., Hartard, F., Ketabdar, H., Möller, S., Modeling user satisfaction with hidden markov model. In: Proceedings of SIGdial (2009)
12. Engelbrecht, K.P., Quade, M., Möller, S.: Analysis of a new simulation approach to dialog system evaluation. Speech Commun. **51**, 1234–1252 (2009)
13. Frostad, K.: Best practices in designing speech interfaces. (2004) http://msdn.microsoft.com/en-us/library/ms994646.aspx
14. Georgila, K., Henderson, J., Lemon, O.: User Simulation for Spoken Dialogue Systems: Learning and Evaluation. In: Proceedings of Interspeech (2006)
15. Gorin, A.L., Riccardi, G., Wright, J.H.: How may I help you? Speech Commun. **23**, 113–127 (1997)
16. Grice, H.P.: Logic and conversation. Syntax Semant. Vol 3. Speech Acts, **3** 41–58 (1975)
17. Hartikainen, M., Salonen, E.P., Markku Turunen, M.: Subjective evaluation of spoken dialogue systems using SERVQUAL method. In: Proceedings of Interspeech (2004)
18. Henderson, J., Lemon, O., Georgila, K.: Hybrid reinforcement/supervised learning for dialogue policies from communicator data. In: Proceedings of the IJCAI workshop on Knowledge and Reasoning in Practical Dialogue Systems (2005)
19. Hirschman, L., Pao, C.: The cost of errors in a spoken language system. In: Proceedings of Eurospeech'93 (1993)
20. Hone, K.S. Graham, R.: Towards a tool for the subjective assessment of speech system interfaces (SASSI). Nat. Lang. Eng. **6**, 303–387 (2000)
21. ITU-T Supplement 24. Parameters describing the interaction with spoken dialogue systems. Technical report, Internationals Telecommuncation Union (2005)
22. ITU-T Rec. P851. 2003. Subjective quality evaluation of telephone services based on spoken dialogue systems. Technical report, Internationals Telecommuncation Union (2003)
23. Janarthanam, S., Lemon, O.: Learning to adapt to unknown users: referring expression generation in spoken dialogue systems. In: Proceedings of ACL '10 (2010)
24. Janarthanam, S., Lemon, O.: A Two-tier User Simulation Model for Reinforcement Learning of Adaptive Referring Expression Generation Policies. In: Proceedings of SIGdial (2009)

25. Kamm, C.: *User Interfaces for voice applications*, pp. 422–442. National Academy Press, Washington, DC, USA (1994)
26. Keeney, R.L., Raiffa, H.: Decisions with multiple objectives: Preferences and value tradeoffs. John Wiley and Sons, New York (1976)
27. Lamel, L., Rosset, S., Gauvain, J.L.: Considerations in the design and evaluation of spoken language dialog systems. In: Proceedings of ICSLP (2000)
28. Levin, E., Pieraccini, R., Eckert, W.: A stochastic model of human-machine interaction for learning dialog strategies. IEEE Trans. Speech. Audio. Process. **8**(1), 11–23 (2000)
29. Lin, B.S., Lee, L.S.: Computer-aided analysis and design for spoken dialogue systems based on quantitative simulations. IEEE Trans. Speech. Audio. Process. **9**(5), 534–548 (2001)
30. López-Cózar, R., Callejas, Z., McTear, M.F.: Testing the performance of spoken dialogue systems by means of an artificially simulated user. Artif. Intell. Rev. **26**(4), 291–323 (2006)
31. Möller, S., Englert, R., Engelbrecht, K., Hafner, V., Anthony Jameson, A., Antti Oulasvirta, A., Raake, E.R., Reithinger, N.: Memo: Towards automatic usability evaluation of spoken dialogue services by user error simulations (2006)
32. Möller, S.: Quality of Telephone-Based Spoken Dialogue Systems. Springer (2005)
33. Möller, S., Ward, N.G.: A framework for model-based evaluation of spoken dialog systems. In: Proceedings of SIGdial (2008)
34. Paek, T., Empirical methods for evaluating dialog systems. In: Proceedings of the Second SIGdial Workshop on Discourse and Dialogue - Volume 16. Association for Computational Linguistics. (2001)
35. Paek, T.: Toward evaluation that leads to best practices: reconciling dialog evaluation in research and industry. In: Proceedings of the Workshop on Bridging the Gap: Academic and Industrial Research in Dialog Technologies, pp. 40–47, Association for Computational Linguistics (2007)
36. Pieraccini, R., Huerta, J.: Where do we go from here? research and commercial spoken dialog systems. In: Proceedings of 6th SIGdial Workshop on Discourse and Dialog, (2005)
37. Pietquin, O.: *A framework for unsupervised learning of dialogue strategies*. Presses univ. de Louvain (2004)
38. Pietquin, O., Hastie, H.: A survey on metrics for the evaluation of user simulations. *Knowledge Engineering Review*, 2013. Accepted for Publication.
39. Putois, G., Young, S., Henderson, J., Lemon, O., Rieser, V., Liu, X., Bretier, P., Laroche, R.: Initial communication architecture and module interface definitions. Technical report, Classic Deliverable D5.1.1 (2008)
40. Rahim, M., Fabbrizio, G.D., Kamm, C., Walker, M., Pokrovsky, A., Ruscitti, P., Levin, E., Lee, S., Syrdal, A., Schlosser, K.: Voice-if: A mixed-initiative spoken dialogue system for. In: Proceedings of Eurospeech (2001)
41. Rieser, V., Lemon, O.: Simulations for learning dialogue strategies. In: Proceedings of Interspeech, Pittsburg (USA) (2006)
42. Rieser, V., Lemon, O.: Reinforcement Learning for Adaptive Dialogue Systems: A Data-driven Methodology for Dialogue Management and Natural Language Generation. Spinger (2011)
43. Rieser, V., Lemon, O.: Automatic learning and evaluation of user-centered objective functions for dialogue system optimisation. In: Proceedings of the Sixth International Language Resources and Evaluation (LREC) (2008)
44. Rieser, V., Lemon, O.: Learning effective multimodal dialogue strategies from wizard-of-oz data: bootstrapping and evaluation (2008)
45. Schatzmann, J., Georgila, K., Young, S.: Quantitative evaluation of user simulation techniques for spoken dialogue systems. In: Proceedings of SIGdial'05 (2005)
46. Scheffler, T., Roller, R., Reithinger, N.: SpeechEval – evaluating spoken dialog systems by user simulation. In: Proceedings of the 6th IJCAI Workshop on Knowledge and Reasoning in Practical Dialogue Systems, Pasadena, CA, USA, pp. 93–98 (2009)
47. Schmitt, A., Schatz, B., Minker, W.: Modeling and Predicting Quality in Spoken Human-Computer Interaction. In: Proceedings of SIGdial (2011)

48. Shriberg, E., Wade, E., Price, P.: Human-machine problem solving using spoken language systems (SLS): factors affecting performance and user satisfaction. In: HLT '91: Proceedings of the workshop on Speech and Natural Language, pp. 49–54. Association for Computational Linguistics (1992)
49. Suendermann, D., Evanini, K., Liscombe, J., Hunter., P, Dayanidhi, K., Pieraccini, R., From Rule-Based to Statistical Grammars: Continuous Improvement of Large-Scale Spoken Dialog Systems, Proceedings of the 2009 IEEE Conference on Acoustics, Speech and Signal Processing (ICASSP 2009), Taipei, Taiwan, April 19–24 (2009)
50. Suendermann, D., Liscombe, J., Pieraccini, R.: Contender. In: Proceedings of the SLT 2010 IEEE Workshop on Spoken Language Technology (2010)
51. Suendermann, D., Liscombe, J., Krishna Dayanidhi, K., Roberto Pieraccini, R.: A handsome set of metrics to measure utterance classification performance in spoken dialog systems. In: Proceedings of SIGdial pp. 349–356 (2009)
52. Walker, M.A., Langkilde-Geary, I., Wright-Hastie, H., Wright, J., Gorin, A.: Automatically training a problematic dialogue predictor for a spoken dialogue system. J. Artif. Intell. Res. **16**, 293–319 (2002)
53. Walker, M., Rudnicky, A., Aberdeen, J., Owen Bratt, E., Garofolo, J., Hastie, H., Le, A., Pellom, B., Potamianos, A., Passonneau, R., Prasad, R., Roukos, S., Greg, S., Stallard, S.D.: Darpa communicator evaluation: Progress from 2000 to 2001. In: Proceedings of ICSLP 02, pp. 273–276 (2002)
54. Walker, M.A., Passonneau, R., Boland. J.E.: Quantitative and qualitative evaluation of DARPA communicator spoken dialogue systems. In: Proceedings of ACL (2001)
55. Walker, M.A., Aberdeen, J., Boland, J., Bratt, E., Garofolo, J., Hirschman, L., Le, A., Lee, S., Narayanan, S., Papineni, K., Pellom, B., Polifroni, J., Potamianos, A., Prabhu, P., Rudnicky, A., Sanders, G., Seneff, S., Stallard, D., Whittaker, S.: Darpa communicator dialog travel planning systems: The june 2000 data collection. In: Proceedings of Eurospeech (2001)
56. Walker, M.A., Rudnicky, A., Aberdeen, J., Bratt, E., Garofolo, J., Hastie, H., Le, A., Pellom, B., Potamianos, A., Passonneau, R., Prasad, R., Roukos, S., Sanders, G., Seneff, S., Stallard, D.: Darpa communicator: Cross-system results for the 2001 evaluation. In: Proceedings of ICSLP (2002)
57. Walker, M.A., Kamm, C.A., Litman, D.J.: Towards Developing General Models of Usability with PARADISE. Nat. Lang. Eng., **6**(3), 363–377 (2000)
58. Walker, M., Passoneau, R.: DATE: A dialogue act tagging scheme for evaluation. In: Proceedings of the Human Language Technology Conference (HLT) (2001)
59. Walker, M.A.: An application of reinforcement learning to dialogue strategy selection in a spoken dialogue system for email. J. Artif. Intell. Res. **12**, 387–416 (2000)
60. Walker, M.A.: Can we talk? methods for evaluation and training of spoken dialogue systems. Lang. Resour. Evaluation **39**(1), 65–75 (2005)
61. Walker, X., Boland, J., Kamm, C.: The utility of elapsed time as a usability metric for spoken dialogue systems. In: Proceedings of the IEEE Automatic Speech Recognition and Understanding Workshop (ASRU99) (1999)
62. Wright-Hastie, H., Prasad, R., Walker, M.: What's the trouble: Automatically identifying problematic dialogues in. In: Proceedings of ACL, pp. 384–391 (2002)
63. Young, S., Gasic, M., Keizer, S., Mairesse, F., Schatzmann, J., Thomson, B., Yu, J.: The hidden information state model: a practical framework for POMDP-based spoken dialogue management. Computer Speech and Language 24(2), 150–174 (2010)

Chapter 8
Data-Driven Methods in Industrial Spoken Dialog Systems

Roberto Pieraccini and David Suendermann

8.1 Introduction

In the early 1990s, the performance of speech and language processing technology combined with advanced voice user interface (VUI) design procedures allowed to start building conversational machines which could be deployed for commercial services offered to a large population of users [11]. Such machines would provide services typically assigned to call centers and human agents or to touch-tone (DTMF) interactive voice response (IVR) systems. Examples include providing travel information for trains or flights, routing phone calls to the appropriate department or agent, performing banking or stock market transactions, and providing technical support and troubleshooting. In general, conversational machines (in the following referred to as spoken dialog systems, or SDSs) consist of the following components (see Fig. 8.1):

- Automatic speech recognition (ASR)
- Spoken language understanding (SLU)
- Dialog manager
- Language generation
- Speech generation, such as text-to-speech (TTS) or, simpler, prompt player

R. Pieraccini (✉)
SpeechCycle, New York, NY, USA

ICSI, Berkeley, CA, USA
e-mail: robertopieraccini@yahoo.com

D. Suendermann
SpeechCycle, New York, NY, USA

DHBW, Stuttgart, Germany
e-mail: david@suendermann.com

O. Lemon and O. Pietquin (eds.), *Data-Driven Methods for Adaptive Spoken Dialogue Systems*, DOI 10.1007/978-1-4614-4803-7_8,

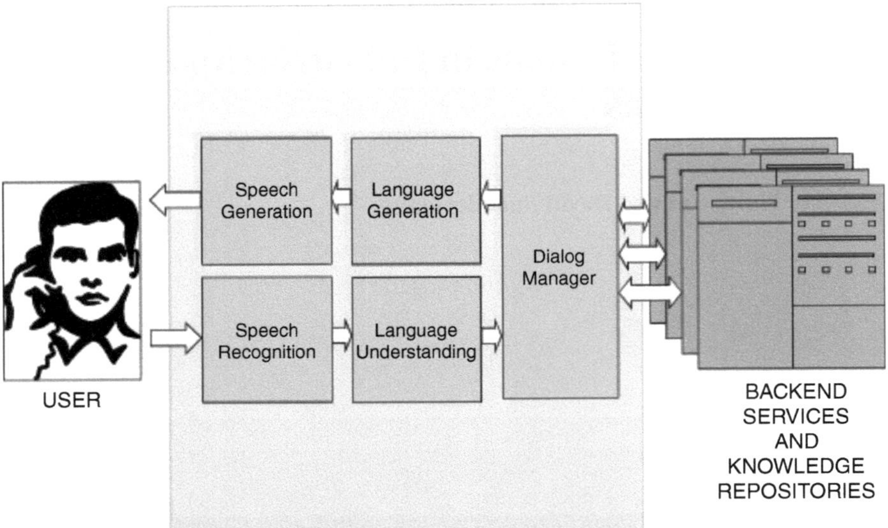

Fig. 8.1 High-level architecture of a spoken dialog system

Industrial interest started to soar, together with the raise of industrial standards, when SDSs demonstrated they could actually (partially) replace costly human agents or touch-tone IVRs in providing simple services, often in a faster and more reliable fashion. The cost saving produced by SDSs is evident when looking at large telecommunication companies, service providers, and enterprises with an extremely high volume of customer service interactions. Applications like How May I Help You (HMIHY) [6], developed by AT&T in the late 1990s, were designed to route callers to the right agent by asking them simple questions to which they could answer, freely, in their own words. Such SDSs can successfully process millions of calls per day allowing for a faster and more accurate response to callers and helping companies save a considerable amount of money. Furthermore, SDSs are easily scalable with incremental or temporary increase of call volume, like in the event of outages, promotions, elections, etc. If these transactions would have to be handled by humans, such situations would result in elevated wait times for callers or the need to hire short-term agents, and that would result in higher costs and a potential decrease of service quality.

Because industrial SDSs are designed to handle large volumes of call traffic, they process large amounts of data on a daily basis, part of which can be stored and used to improve the systems themselves. This data includes:

- Full-duplex call recordings (2,355 kB)
- Individual speech utterances as processed by the speech recognizer (277 kB)
- Speech recognition logs (76 kB)
- Voice browser logs (351 kB)
- Application logs (106 kB)

We report above, in parentheses, the average per-call size for each item calculated on one of SpeechCycle's Internet troubleshooting applications deployed for a large US provider [1]. Considering that this application handles about 500,000 calls every month, to store the data for all the calls handled by the system would require about 1.5 TB storage per month.

The availability of such large amounts of data poses the question of how to effectively use it for improving the performance of the system at hand. In this chapter, we describe several methods we effectively implemented with this goal in mind.

The most obvious data-driven optimization is related to speech recognition accuracy. Significant improvements in this domain can be achieved efficiently and effectively by using statistical grammars—rather than rule-based ones—and by creating an automated procedure of data collection, transcription, and semantic annotation, as well as grammar learning and optimization.

However, besides the accuracy of the speech recognizer, in an SDS, there are many other issues that determine the overall performance. In particular, we discuss three other data-driven optimization problems: optimizing the escalation decision (i.e., when to transfer a call to a human agent), finding what the best order for a sequence of questions is, and optimizing the use of alternative VUI strategies at different points in the dialog.

8.2 Data-Driven Grammar Optimization

As is common knowledge in the speech recognition industry (and in the general population), one of the most evident flaws of a SDSs can be directly related to inaccurate results produced by the speech recognizer. Industrial SDSs typically use rule-based grammars (based on the SRGS[1] standard) to represent the set of possible words or phrases that can be uttered by users at any specific state of the dialog when directed to speak a precise word or phrase by a system prompt. SRGS grammars also allow associating a number of semantic tags, or even scripts, with any expression described by the grammar. By restricting the scope of the speech recognizer search space, rule-based grammars can achieve very good performance for in-grammar utterances, that is, when the user speaks exactly as instructed by the prompt (i.e., *directed-dialog mode*). However, we know that users do not always comply with the directions of the prompt but can produce a large variety of expressions that are hard to predict and thus to account for in the grammar at design time. Rule-based grammars have poor, or no, generalization ability and have to be updated mostly by hand. Generally, industrial systems go through regular tuning cycles where expert VUI designers and speech recognition specialists (also known as *speech scientists* in the industry) analyze the speech recognition performance and the data collected

[1] Speech recognition grammar specification [7].

during the deployment and modify grammars in order to increase the in-grammar accuracy while reducing the out-of-grammar rate. This is a costly and laborious activity.

Statistical language models or SLMs (usually based on n-grams) complemented by statistical semantic classifiers (à la HMIHY) have been typically used in dialog turns where an open prompt (e.g., *please tell me the reason you are calling about*) leaves users the freedom to express what they want in their own words (i.e., *natural language dialog mode*). However, the expertise, the tools, and, most important, the data needed to create an SLM and a semantic classifier are not typically available at the design stage of an industrial SDS. Moreover, there has been a popular perception in the industry that statistical grammars (i.e., SLMs including semantic classifiers), although useful in natural language dialog situations, cannot perform as well as rule-based grammars in directed dialog. This belief does not take into account the fact that users, more often than not, speak out-of-grammar, and tuning rule-based grammars by hand is a costly exercise. The other reason, besides the lack of data, that encourages the use of rule-based grammars in industrial deployments, for all the directed-dialog turns, is that rule-based grammars are human readable and can be easily modified by a developer, while statistical grammars can consist of thousands of floating point values that represent probability models making little sense to nonspecialists and that are not supposed to be modified by hand.

Having deployed several industrial dialog systems which generate large amounts of speech data, several years ago, we embarked into a project aimed at substituting all rule-based grammars in directed-dialog turns—hundreds of grammars are typically used in a deployed SDS—and created an automated workflow which would, regularly, update the statistical grammars when enough data is available for each particular turn. This process, which autonomously releases new grammars for deployment when they show a statistically significant improvement in accuracy over the ones in production, showed to greatly outperform any manual tuning based on rule-based grammars [13].

In order to be able to assess the performance of grammars, we need first to establish an accuracy measure that takes into account phenomena such as out-of-grammar utterances and rejection, which are common in an SDS. This measure, which we call *true total* (or *true confirm total* if confirmation is taken into account), is based on a classification of possible speech recognition events as described by the chart in Fig. 8.2. At each turn of the dialog, an utterance can be classified as *in-grammar* or *out-of-grammar*, which, in the most general case, depends on whether the semantics of the utterance is within the scope of the grammar. In either case, the speech recognizer can *accept* the result or *reject* it on the basis of a confidence measure. In case of in-grammar utterances, the accepted result can be *correct*—i.e., corresponding to what the user said—or *wrong*. An accepted out-of-grammar utterance is always wrong by definition. We can then consider the possibility of asking the user for confirmation (e.g., *I think you said* Austin, *is that correct?*) of the accepted speech recognition result, based on a confidence threshold. If we do not take into account confirmation, the true total measure characterizes the rate at which the recognizer makes *good decisions*, in other words, the overall rate of

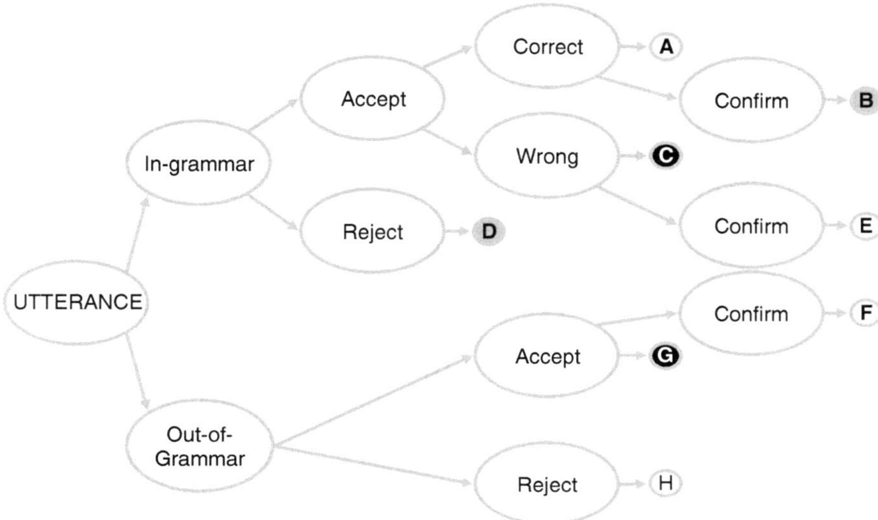

Fig. 8.2 Classification of possible speech recognition events in a spoken dialog system

Table 8.1 Improvement in recognition accuracy after replacing rule-based grammars with statistical ones

Number of calls	533,347
Number of utterances	2,184,203
Number of grammars	145
Original true total	77.9%
True total with statistical grammars	90.5%

correctly accepted in-grammar utterances plus rejected out-of-grammar ones, over all processed utterances (nodes A+H in the chart of Fig. 8.2). Notice that the events identified by gray nodes in the chart of Fig. 8.2 are mildly severe events for a dialog system, while those identified by black nodes are considered severe and can disrupt the course of the interaction.

When we started converting all rule-based grammars in our deployed systems to statistical grammars, we reported significant improvement in the accuracy of the speech recognition process (i.e., significantly higher values of true total). As an example (Table 8.1), we replaced most of the grammars—at least those for which we had enough data—in a technical support application with equivalent statistical grammars derived from more than half a million calls (533,347) corresponding to 2.2 million utterances. The total number of grammars was 145. The original overall accuracy (true total) was 77.9%. After the statistical grammars were deployed, we measured an overall true total of 90.5% [13].

8.3 Stay or Leave? Optimizing the Escalation Decision

The ability to decide when it is the right time to escalate a call to a human agent is one of the crucial elements responsible for the overall automation rate and the quality of the caller experience in an industrial SDS. It is often a trade-off decision, especially considering the rewards and costs at stake, besides the potential frustration of the users. In fact, quantifying the reward (or the costs, i.e., the negative reward) of an SDS interaction is the starting point of any type of optimization.

Every call processed by an SDS is subject to rewards and costs. From the point of view of the enterprise that deploys an SDS, a successfully automated call corresponds to a certain amount of savings when compared to the same call handled by a human agent—human agents are typically more expensive than automated SDSs. Correspondingly, from the point of view of the vendor who builds, maintains, and hosts the SDS, a successfully automated call brings a certain amount of revenue. These savings, or revenue, can be quantified on a per-call basis. Moreover, there are costs which are based on the duration of the call, such as telecommunication, licensing, and application hosting costs. Thus, at any point during an automated dialog, a trade-off decision should be taken on whether to continue the call or escalate it to a human agent. If the system escalates a call that would, eventually, have been successfully automated, you simply cannot realize the potential rewards, savings, or gains that call would have produced. If the system insists in trying to automate a call that ends up being escalated, you incur additional per-minute costs without reaping the benefits of a successful automation. So, the best strategy would be to escalate a call that most likely will not be successfully automated as soon as possible. With that, you reduce the costs, and, even more important, you reduce the user frustration, without forcing them to go through a long and unsuccessful call just to be escalated to a human agent, anyway, at the end.

In one of the first experiments [9], we quantified the costs with two matrices (Fig. 8.3), one representing the per-call costs (C^N) and one representing the per-minute costs (C^T).

For instance, c^N_{NC} represents the cost for a call which is eventually nonautomated (i.e., the caller's problem has not been solved) but continued to the very end (i.e., the call terminates without an escalation to an agent), while c^T_{NE} is the *per-minute* cost of a call which is not automated and terminated by a transfer (escalation) to an agent. These costs can be computed according to the particular business model (e.g., the price paid by a customer for an automated call or the equivalent saving in agent's time) and the actual costs associated with running the automated system (e.g., the per-minute telecommunication and hosting charges). The specific values of the elements of C^N and C^T determine the optimal escalation strategy and the overall level of reward (or savings) once it is implemented.

In one of the experiments, we set c^N_{AC} to -1 (i.e., a reward of 1 for each automated call), c^T_{AC} and c^T_{NC} to 0.1 (i.e., a cost of 0.1 per minute for each non-escalated call regardless of its level of automation), and all the other values to 0. We then assumed a decision point, based on an *automatable/nonautomatable* (*A*/*N*) prediction to

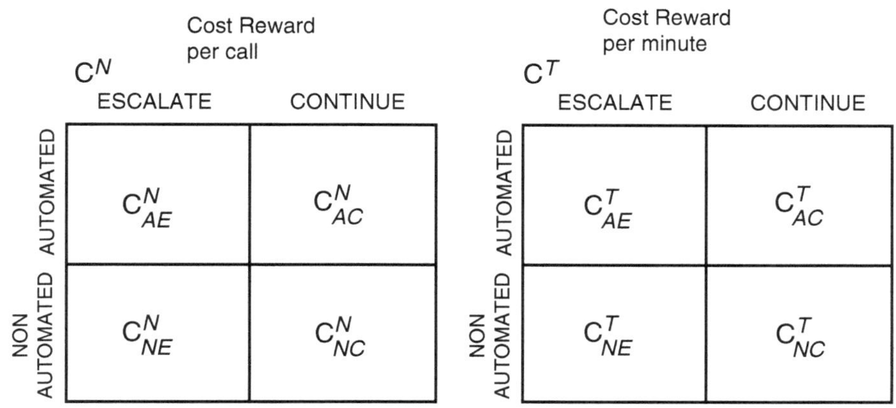

Fig. 8.3 Cost parameterization for escalation decisions

be set early in a call, when enough information has been gathered to be able to comfortably predict the outcome of the call. With that choice of costs, the cost of each one of the two actions (i.e., escalation or continue the call) is

$$C_E = 0; \quad C_C = \begin{cases} 0.1d - 1 & \text{if } A, \\ 0.1d & \text{if } N, \end{cases} \tag{8.1}$$

where d is the duration of the call in question after the decision point. Thus, in order to guarantee that the expected reward is a positive value, a call classified as *nonautomatable* should be escalated if

$$p(N|n)\bar{d}c^T_{NC} + p(A|n)(\bar{d}c^T_{AC} + c^N_{AC}) > 0, \tag{8.2}$$

where $p(N|n)$ is the probability of a nonautomated (n) call to be classified as nonautomatable (N)—in other words, the *precision* of the classifier—and $p(A|n)$ the probability of a nonautomated call to be classified as automatable (A). Thus, the cost saving achieved by implementing the strategy of Eq. (8.2) can be computed as

$$\bar{S} = (\bar{d}c^T_{NE} + (1 - p(N|n))c^N_{AC})p(n), \tag{8.3}$$

where $p(n)$ is the prior probability of a call to be nonautomated. Thus, the precision (and, in general, the recall) of the classifier determines the strategy and the level of savings.

In the experimental evaluation, we used a corpus of 61,531 call logs from a telephone-based technical support system deployed for a large cable TV company. The system was designed to help callers solve typical problems occurring in a residential cable TV setting, such as noisy or absent picture or problems with on-demand channels. The full automation rate was about 23%; the rest of the calls were

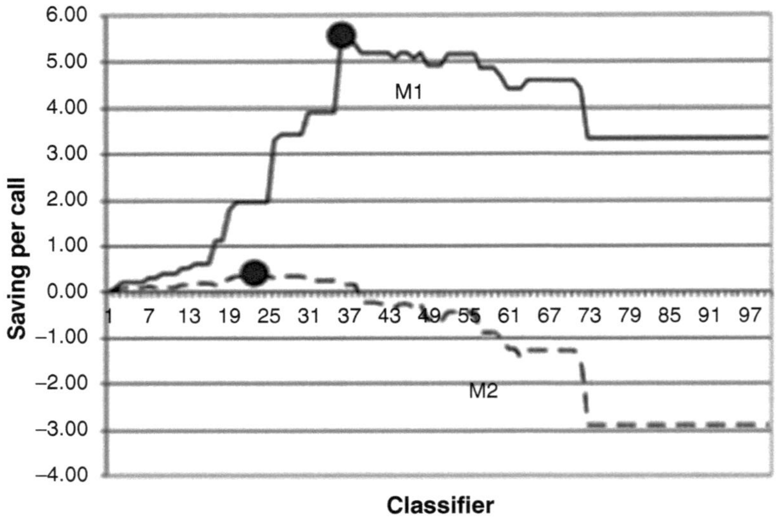

Fig. 8.4 Saving achieved by each classifier for two different cost structures

either abandoned by the caller or escalated. 15,000 of the 61,531 calls were used for development and 10,000 for test; all the remaining calls were used for training. We then implemented a number of automatable/nonautomatable classifiers based on SLIPPER [3], each one of them with a different figure of precision and recall. The features used for classification, collected on the initial part of the dialog up to the call reason identification state, were of the following types:

- **Speech recognition features**: e.g., recognition status (rejection, no-input, recognition), recognized word string, utterance duration, number of words, recognition confidence, DTMF input.
- **Language understanding features**: e.g., SLU hypotheses and corresponding confidence measures.
- **Dialog features**: e.g., cumulative counts of re-prompts, confirmations.

All features were extracted automatically, and each call was automatically labeled as A or N. We estimated 100 classifiers by running SLIPPER on the training data with different weighting of the two classes of training samples (A and N). Consequently, each classifier showed a different figure of precision/recall and thus a different performance with respect to the expected saving. The results can be summarized by Fig. 8.4 where the saving per call for each of the 100 classifiers is reported for two different cost structures (savings here are represented as cents of the cost unit).

The curve tagged as $M1$ represents the previously introduced costs (0.1 per-minute costs and 1.0 saving per automated call) and $M2$ an alternative cost schema with 0.05 per-minute costs and 0.7 saving per automated call. The filled circles in the figure represent the best classifier for each cost structure and the actual saving which is 5.7 cents per call for the $M1$ cost structure [9].

The seemingly minimal savings induced by this method can be attributed mainly to two reasons. The first is that, early in a call, there is not enough evidence on whether the call will be automated or not. Second, clearly the ability to automate a call depends on the call reason which is indeed taken into account in the described method. However, whether a call gets automated or not is also influenced by the resolution path taken by the system later in the dialog which may depend on other factors than just the call reason and is unknown at the time the *A/N* classification is performed. These considerations led us to develop another method, described in the next section, which is simpler and more effective than the one described in this section.

8.4 State-Dependent Escalation Through Minimally Invasive Call Flow Surgery

The call flow of a commercial dialog system is a step-by-step finite-state representation of the actions taken by the dialog system for each potential user input and can be quite complex. It is not unusual to design, build, and maintain call flows which include thousands of nodes, tens of thousands of prompts, and hundreds of grammars. During the lifetime of a deployed dialog system, the accumulated log data can give information about which branches of the call flow provide better chances of automation and which others do not. The main idea of the technique described in this section is that of analyzing the call flow, node by node, and determining the potential of automation for any node reached by a call. Then, the call flow is *pruned* by identifying those nodes with a low automation potential and restructured by associating an escalation procedure with those nodes, thus eliminating any further call flow path following them.

In order to perform an effective pruning of a call flow, and thus escalate a call when a node with a low chance of automation is traversed, we need to define a proper *reward* function which would allow ranking the nodes in terms of the average reward of all the calls going through them. The first consideration is that the reward should be proportional to the savings achieved by a set of automated calls as compared to the situation in which human agents would handle the same calls. Hence, the average saving for this set of calls can be calculated as

$$S = W_A A - W_T T, \tag{8.4}$$

where W_A is the average cost of a call handled by a human agent, T is the average duration of a call handled by the automated SDSs, W_T is the associated cost per minute (telecom, hosting, etc.), and A is the effective automation rate of the system (percent of calls which are resolved without the caller being escalated or calling back after hang-up). In this experiment, we represent the savings in seconds (rather than in a currency value), that is, we use the following reward measure:

$$R = T_A A - T \tag{8.5}$$

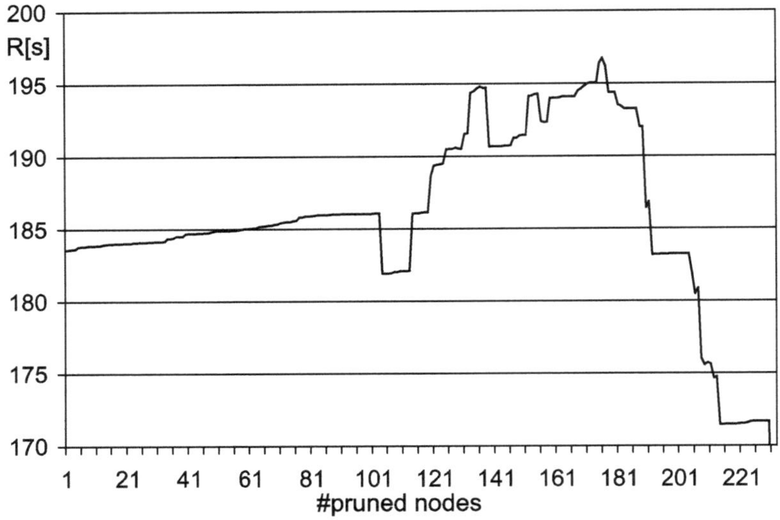

Fig. 8.5 Value of the average reward obtained by removing low-ranking nodes, one at a time

with

$$T_A = \frac{W_A}{W_T}. \tag{8.6}$$

Intuitively, R is proportional to the equivalent human agent time saved by using the automated system, and T_A is a normalization factor.

We analyzed 45,631 call logs. Collectively, the calls visited 847 nodes of the call flow. The average reward, computed according to Eq. (8.5), was 183.5 s, using a T_A value of 5,000 s which was estimated using the parameters of the actual business model of the deployed system. We then ranked all of the 847 visited nodes with respect to the average reward of all the calls that traversed them. Notice that each call visited a subset of the 847 nodes, which is a subset of the total number of nodes of the call flow, some nodes of which were never visited by any call in the set. It is expected that most of the calls visit those nodes that are located early in the dialog (i.e., close to the root of the call flow), while nodes close to the leaves of the call flow are visited only by a few calls. After all of the visited nodes were ranked according to the R value, we started removing (pruning) the nodes with the lower rank, simulating a strategy that would escalate a call at each removed node. The chart of Fig. 8.5 shows the average reward R obtained after removal of each node, one at a time, according to their rank.

By removing more and more nodes (up to 176), and consequently escalating the calls reaching these nodes, the reward increases almost monotonically. This is due to the reduction of the call time (i.e., T in Eq. (8.5)) without appreciable reduction of the automation A. This means that the removed nodes, and thus the call flow paths following them, did not appreciably contribute to the automation. Instead, callers were kept engaged for a long time without effectively solving their problem and thus frustrating them, thereby reducing the quality of the caller experience. After removing more than 176 nodes, we notice a dramatic drop in automation rate. In other words, these nodes are those that mostly contribute to the automation, and by removing them (i.e., escalating when a call reaches one of them), we miss concrete opportunities of automated calls. The resulting optimal reward, after pruning the lowest-ranking 176 nodes, is 196.8 s, with a net gain of $196.8 - 183.5 = 13.3$ s compared to the original call flow [14].

8.5 Asking the Right Question at the Right Time

Often, a SDSs needs to ask a number of questions to the caller in order to determine the identity of a piece of information. Let us take for instance the problem of determining which type of modem a caller has at home. Consider the case where there are only three types of modems that can be identified as follows:

1. Black Ambit
2. White Ambit
3. Black Arris

Two questions considered by the VUI designer to distinguish between these three types of modems are:

(A) *Is your modem black or white?*
(B) *Do you have an Ambit or an Arris modem?*

We notice that when question A is asked first, and the response is *white*, we know that the modem is of type 2, and question B can be skipped. Conversely, when question B is asked first, and the response is *Arris*, the modem is of type 3, and question A does not have to be asked. So, apparently, the order in which questions are asked in a call flow (and, generally, the order in which activities are carried out in a call flow) can make a difference.

If the distribution of the three modems across the population of callers is uniform, the order in which the questions are asked is irrelevant since both special cases (type 2 suggesting question order A→B and type 3 suggesting order B→A) are equally likely with $p = \frac{1}{3}$. The average number of questions asked in a large number of calls would therefore be $p + 2 \cdot (1 - p) = \frac{5}{3}$. However, if the distribution of modems is not uniform, the order in which questions are asked does make a difference. Think, for instance, of the case where the vast majority of the population has a white Ambit, thus asking the question about the color of the modem first would make the average number of questions asked very close to 1.

A well-designed system should ask the *most informative* question first, in other words, that question that, on average, leads to the identification of the target in the least number of steps. In mathematical terms, this concept is quantified by the notion of *information gain*, as defined in [10]. The information gain of the random variable D (the destination, e.g., the set of all possible modems $d_1,\ldots,d_I$) for a specific question Q (whose possible answers are $q_1,\ldots,g_J$, e.g., the different possible colors of a modem) can be computed as

$$I(D;Q) = H(D) - H(D|Q) \tag{8.7}$$

with Shannon's entropy

$$H(D) = -\sum_{i=1}^{I} p(d_i)\log_2 p(d_i) \tag{8.8}$$

and the conditional entropy

$$H(D|Q) = -\sum_{j=1}^{J} p(q_j)\sum_{i=1}^{I} p(d_i|q_j)\log_2 p(d_i|q_j). \tag{8.9}$$

The optimal strategy in order to minimize the average number of questions consists thus in asking the question, at each step, that results in the highest information gain and that had not been asked before. Prerequisite in order to apply this approach is to have knowledge about the involved probabilities $p(d_i)$, $p(q_j)$, and $p(d_i|q_j)$ for each question, answer, and destination. While these probabilities can be based on rough estimates at the initial design stage of an application (e.g., drawn from call center statistics or similar knowledge sources), after deployment to production, they can be computed using actual logs collected for production calls.

We tested this algorithm in a deployed call router for a large cable company that processes four million calls per month routing callers to one of 20 possible destinations. The application can ask questions about the following attributes:

- The type of service requested by the caller (e.g. orders, billing, technical support)
- The product for which the service is requested (e.g. Internet, cable TV, telephone)
- The action requested (e.g. cancel a service, schedule an appointment, make a payment)
- Additional attributes (e.g. credit card, pay-per-view, digital TV conversion)

We analyzed 3.9 million calls that required an average of four questions to identify the correct destination. After applying the algorithm described here, the average number of questions dropped to 2.9, which translates into a substantial reduction of the average handling time and a respective gain of the savings induced by the call router [16].

8.6 Data-Driven Optimization of the Dialog Strategy

There has been, and currently there is, a lot of work on using machine learning for estimating and optimizing the strategy used by dialog systems (see Chap. 5). The first work in this area started in the late 1990s with the representation of a dialog system as a Markov decision process (MDP) [8] and the use of reinforcement learning to derive the optimal policy (or strategy). Levin et al. [8] demonstrated that it is indeed possible to derive a strategy solely by defining a finite set of actions (e.g., prompts, database access), a state space (e.g., a set of variables representing the current state of the interaction), and a reward function, and then to use a reinforcement learning algorithm while the system is interacting with users. Starting with a randomly selected strategy, the reinforcement learning algorithm determines the optimal action for each state maximizing the expected reward on a finite or infinite window of time. A refinement to MDPs is partially observable Markov decision processes (or POMDPs) [18] that model the uncertainty inherent to SDSs, particularly that derived from potential errors of speech recognition and language understanding.

The problem with learning dialog strategies while interacting with users is that the estimation of an optimal or even a reasonably suboptimal strategy may take a large number of interactions during which the system may behave haphazardly. Using real users for the training of a dialog strategy is thus not feasible for a commercial system and often too expensive for a research system (recent advances in academic research to address this issue involve sample-efficient algorithms using, e.g., Gaussian processes [5] or batch learning [12]). This is why *simulated* (or *artificial*) users were developed since the early days of data-driven methods for SDSs [4]. As discussed in Chap. 4, a simulated user is realized by a dialog system that behaves like a user: It responds to prompts within a range of answers sampled from the actual statistical distribution observed during a data collection phase. Simulated users are typically realized at the semantic level, that is, they generate semantic tags and not linguistic expressions. If, at a certain state of the dialog, the system's action consists in asking a specific yes/no question, the simulated user randomly generates a *yes* or a *no* or any other possible semantic tag. This generation is based on a probability distribution reflecting the behavior of that particular dialog state drawn from a known population of users.

Commercial systems are too complex and too constrained by business and customer requirements to lend themselves to a complete estimation of the strategy based on machine learning, at least with the currently available techniques. However, we can effectively use simpler learning techniques to optimize the performance of a dialog system with respect to a number of potential alternative designs. To that end, we developed a technique called *Contender* [15].

We start with the consideration that the activity associated with a certain node of a call flow could be implemented by a number of alternative designs. Consider, for instance, a prompt requesting the reason for the call. One could use a general open prompt (e.g., *Please tell me the reason you are calling about*) or a menu (e.g., *Please say* technical support, account information, *or* pay bill) or even a multistep strategy (e.g., *This is the technical support hotline. If you don't need technical support, say* other). A more sophisticated example can be that of alternative strategies for fixing the same problem, for instance, lack of Internet connectivity in a digital service provider support application. For instance, one strategy could be sending a reboot signal to the modem, while the other strategy could be asking the caller to power cycle it. In different situations, experienced VUI designers can come up with alternative designs, like in the above examples. However, it is not until the system is deployed and tested that we can determine the effectiveness of the chosen alternative with the specific population of users. If the chosen design does not produce the expected results, the designer implements one of the other alternatives in a new release of the system which is then deployed and tested. This is a very elaborate procedure that may require a large amount of time before coming to an optimal design. Moreover, the changes that may occur in different parts of the call flow between two consecutive releases and the changes that may occur in the environment (e.g., different populations of callers, new products offered by the provider) make it difficult to come to a fair comparison of the different designs. The situation is ever more complex when several parts of the call flow are tested for different designs. Moreover, we also need to consider the fact that some of the designs may be optimal for certain conditions and not for others (e.g., callers having different devices, for instance, analog or digital cable boxes, for a cable TV provider application). All of these considerations make it hard, or even impossible, to test different designs in a commercial dialog system, and the choice is generally left to the experience and intuition of the designer.

Whether variations of the wording of a single prompt can produce significant changes of the automation rate, let alone more complex variations of troubleshooting strategy or even speech recognition error-correction strategies, was shown in an earlier experiment discussed in [1]. The problem was that of the design of an initial prompt for the call router deployed at a large digital service provider. At the time, the designers came up with three potential choices:

- Original: An open prompt such as *Please tell me the reason you are calling about*
- With examples: An open prompt with examples, such as: *Please tell me the reason you are calling about, for instance* I have no service
- Offer choices: An open prompt with a choice option, for instance, *Please tell me the reason you are calling about, or say* what are my choices

Figure 8.6 shows the cumulative average deflection (i.e., automation rate) for each prompt, obtained by maintaining averages of all the calls using that prompt. For each incoming call, one of the three prompts was chosen at random; the three options were equiprobable. The chart shows the automation rate computed over a period of two weeks, while receiving approximately 1,500 calls per day. The experiment was terminated when we deemed that the differences among the performance of the three alternatives were statistically significant. At that moment, we estimated a difference of approximately 0.8% points between the best prompt (with examples)

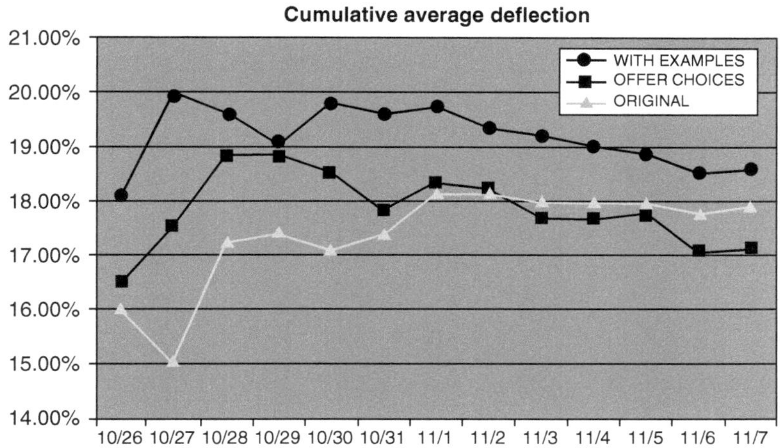

Fig. 8.6 Cumulative average deflection (i.e., automation rate) for a call router with three different opening prompts

and the second best (original). Then, we decided to issue a new release of the system using only the best performing prompt.

Similarly to the experiment described above, the Contender technique allows instrumenting a call flow in such a way that several alternative designs can be associated with one or more nodes of the call flow (Contender nodes). When a call traverses a Contender node, one of the alternatives is randomly selected according to a probability schema characteristic of that particular node. Generally, the selection of alternative designs is set to be equiprobable at the beginning of a Contender experiment. As the system keeps handling calls, each one of them going through a random selection of one of the alternative designs, information regarding the associated call flow paths is logged so that, for each call, we know exactly which choices have been selected at each Contender node. At regular intervals, the logs are analyzed, and the probabilities associated with the alternatives at each Contender node are modified in order to favor the selections which provided the highest value for the average reward. In general, the probabilities are modified in order to satisfy an *exploration-exploitation* criterion. None of the alternatives is ever set to zero probability even though their automation was clearly and in a statistically significant manner lower than the other alternatives. This is done to account for possible changes in the population behavior. Figure 8.7 shows a functional sketch of the Contender process.

The question of how often one should update the probabilities arises when considering that, in industrial settings, any modification to a production system may require the involvement of stakeholders from project management, VUI design, engineering, quality assurance, and operations, and thus there are costs associated with any system change. A reasonable strategy would be that of performing the probability update, for each individual Contender node, when the differences among the performance of the alternative designs reach statistical significance. However,

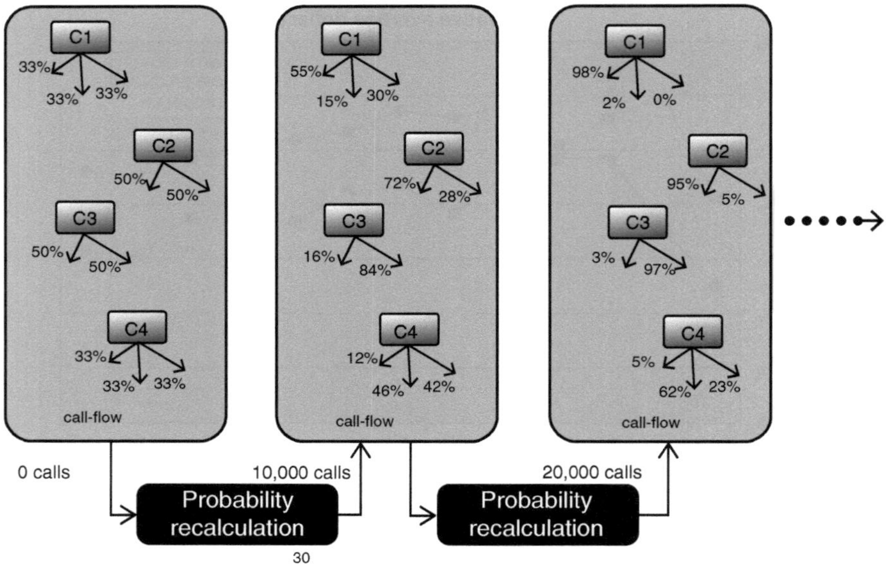

Fig. 8.7 Schematic example of the probability update process of Contender nodes in a call flow

that could take a long time and would prevent reaping the benefits of the better designs early on.

In [15], we developed a probability update schema which outperforms, in terms of cumulated reward, that of choosing the optimal design when the reward difference among alternatives is statistically significant. In practice, that consists in assigning to each alternative branch of a Contender node a probability which is equal to the probability of that branch to be the *winner*, that is, to be selected as the best design when statistical significance is reached.

Contenders proved to be an effective tool not only to boost the performance of SDSs but, particularly, as a handy instrument for VUI designers. Up until the introduction of Contenders, call flow design was mainly based on "best design practices" as communicated by expert designers in text books (e.g., [2]), at industrial conferences (such as the annual SpeechTEK event), and through other publication channels. Such best practices give guidance on:

- What type of question is best in which context (directed, open-ended, yes/no).
- What are good prompts for announcement, question, confirmation, no match, retry, etc.
- Which settings should be used for rejection/confirmation thresholds, recognizer sensitivity, background music volume, etc.
- What are successful strategies to retrieve the call reason, to prevent callers from asking for an agent, or for troubleshooting technical issues, and so on.

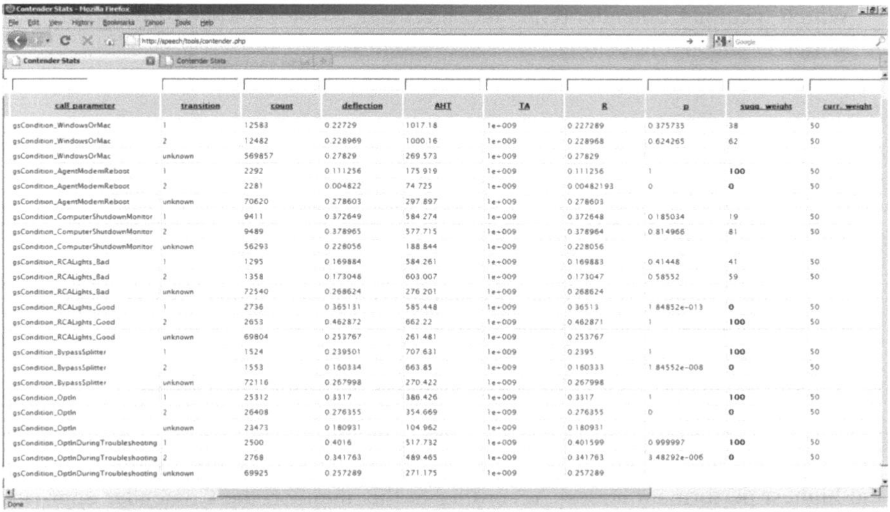

call parameter	transition	count	deflection	AHT	TA	R	p	sugg. weight	curr. weight
gsCondition_WindowsOrMac	1	12583	0.22729	1017.18	1e+009	0.227289	0.375735	38	50
gsCondition_WindowsOrMac	2	12482	0.228969	1000.16	1e+009	0.228968	0.624265	62	50
gsCondition_WindowsOrMac	unknown	569857	0.27829	269.573	1e+009	0.27829			
gsCondition_AgentModemReboot	1	2292	0.111256	175.919	1e+009	0.111256	1	**100**	50
gsCondition_AgentModemReboot	2	2281	0.004822	74.725	1e+009	0.00482193	0	**0**	50
gsCondition_AgentModemReboot	unknown	70620	0.278603	297.897	1e+009	0.278603			
gsCondition_ComputerShutdownMonitor	1	9411	0.372649	584.274	1e+009	0.372648	0.185034	19	50
gsCondition_ComputerShutdownMonitor	2	9489	0.378965	577.715	1e+009	0.378964	0.814966	81	50
gsCondition_ComputerShutdownMonitor	unknown	56293	0.228056	188.844	1e+009	0.228056			
gsCondition_RCALights_Bad	1	1295	0.169884	584.261	1e+009	0.169883	0.41448	41	50
gsCondition_RCALights_Bad	2	1358	0.173048	603.007	1e+009	0.173047	0.58552	59	50
gsCondition_RCALights_Bad	unknown	72540	0.268624	276.201	1e+009	0.268624			
gsCondition_RCALights_Good	1	2736	0.365131	585.448	1e+009	0.36513	1.84852e-013	**0**	50
gsCondition_RCALights_Good	2	2653	0.462872	662.22	1e+009	0.462871	1	**100**	50
gsCondition_RCALights_Good	unknown	69804	0.253767	261.481	1e+009	0.253767			
gsCondition_BypassSplitter	1	1524	0.239501	707.631	1e+009	0.2395	1	**100**	50
gsCondition_BypassSplitter	2	1553	0.160334	663.85	1e+009	0.160333	1.84552e-008	**0**	50
gsCondition_BypassSplitter	unknown	72116	0.267998	270.422	1e+009	0.267998			
gsCondition_OptIn	1	25312	0.3317	386.426	1e+009	0.3317	1	**100**	50
gsCondition_OptIn	2	26408	0.276355	354.669	1e+009	0.276355	0	**0**	50
gsCondition_OptIn	unknown	23473	0.180931	104.962	1e+009	0.180931			
gsCondition_OptInDuringTroubleshooting	1	2500	0.4016	517.732	1e+009	0.401599	0.999997	**100**	50
gsCondition_OptInDuringTroubleshooting	2	2768	0.341763	489.465	1e+009	0.341763	3.48292e-006	**0**	50
gsCondition_OptInDuringTroubleshooting	unknown	69925	0.257289	271.175	1e+009	0.257289			

Fig. 8.8 Portal for calibration and retrieving statistics of Contenders deployed in production SDSs

Many of these recommendations turn out to be non-generalizable, i.e., the optimal choice can depend on the minutiæ of the application, caller population, or other parameters of the call flow at hand. Moreover, there are many questions raised in the design phase of a call flow that are so specific to the application that they have not been discussed elsewhere. For instance, in cable TV troubleshooting, is it better to just ask callers to unplug their cable from the back of the box or to also mention that they can alternatively do it from the wall?

Contenders can be leveraged to answer arbitrary questions coming up during call flow design by implementing all the alternatives deemed reasonable as separate Contender nodes. Due to the aforementioned complexity of industrial call flows, there are numerous places subject to design alternatives; accordingly, multiple Contenders may have to be implemented in the same application. Alternative designs, however, go along with even more complex call flows. Furthermore, we explained above that Contender analysis involves determining winning (randomization) probabilities which potentially differ depending on which geographical area the application is deployed to, whether callers are private or business customers, and other factors. This adds yet another level of complexity to Contender-assisted design which suggests the establishment of a centralized portal for managing all the Contenders available in production deployment. Figure 8.8 shows an example of such a portal managing more than 150 Contenders deployed to SpeechCycle's applications. The total number of calls reported in the portal to date amounts to almost 150 million. Among others, the portal reports about:

- Product, customer, and population the Contender applies to (e.g., Internet troubleshooting for a certain cable provider in Brooklyn, New York)

- Which specific Contender (e.g., `gsCondition_WindowsOrMac`)
- Which Contender node (e.g., `1`, `2`, `3`, or `unknown`)
- The number of calls having traversed this node
- A
- T
- T_A
- R
- The suggested probability per node
- The current probability
- The number of automated calls per month gained by running the Contender
- The duration the Contender was active
- Time stamps of when the last probability analysis took place
- A complete history of all Contender analyses performed in the past

One way to show the effectiveness of the Contender approach is to estimate the additional number of automated calls gained by systematically routing traffic to the highest performant nodes. According to this number, the Contender approach currently increases the number of automated calls processed by SpeechCycle's applications by over 100,000 which translates into a significant increase in revenue [17].

8.7 Spoken Dialog Systems in the Age of Smartphones

As the traditional landline telephone is rapidly and massively giving way to wireless smartphones, a new paradigm for human–computer spoken dialog is appearing. The fact that smartphones are constantly connected to the Internet, are always available, carry substantial computational power, and have a rich display is changing the way speech recognition is deployed for a variety of applications. During the past few years, we saw a few of those applications being used in high volumes with popular platforms such as the iPhone and Android. These applications mainly allow smartphone users to perform searches on the Web by voice, to dial out by simply speaking a name or a number, or to invoke applications, such as the calendar, by just speaking a command. These speech recognition applications have a distributed architecture. Speech is collected, end-pointed, digitized, and compressed on the device by a client generally embedded in the application. One of the advantages of having the incoming speech collected and digitized on the device is the increased signal quality—speech can be collected at a higher bandwidth than typical telephone speech, and local noise cancelation can be performed, e.g., using multiple microphones on the device. The advantage of running speech recognition on a remote server is the extensively more powerful computation and storage capacity than available on the device.

Once collected, the compressed speech waveform is sent to a remote speech recognition server (often referred to as *cloud-based* speech recognizer) via the smartphone's data link (typically using HTTP). Then, after recognition, the text string representing the transcription of the user's utterance—or the set of text strings representing the n-best transcriptions—is sent back to the user's smartphone and the invoking application. For instance, in a voice search application, the text string representing the transcription of the user's spoken query feeds directly a search text box, and the search result is finally displayed.

Bare voice search does not need any form of language understanding since the search engine requires only the textual transcription of the spoken query. Other applications need some type of semantic analysis or information extraction before the input from the recognizer can be used. For instance, in a voice dialing application, the contact name and the type of telephone (e.g., cell, work, home) or the telephone number have to be extracted from the string. For more complex command-and-control applications, semantic classification or parsing is required.

One application that created a large wave of popularity and a revived interest in speech recognition by the consumer population is Siri which was embedded in a new version of the iPhone released at the end of 2011. Siri is a personal agent; it knows about things like your personal contacts and your appointments and has access to popular Web services like restaurant and flight reservation. Siri uses both voice input and output, has a catchy personality, and can respond, in a funny and witty way, to requests such as "What's the meaning of life?" or "What do you look like?" However amazing or advanced Siri looks like to the general population, it does not seem to use any strikingly different technology than that deployed in other commercial dialog systems. Although Siri's architecture has not been publicly disclosed as of yet, it is claimed to use a commercial speech recognition engine and some form of grammar-based language understanding. Hence, the novelty brought by Siri is not primarily algorithmic, but technological. Siri constitutes a well-executed piece of engineering, seamlessly integrated with the rest of the iPhone software.

8.8 Conclusions

In this chapter, we demonstrated how a variety of data-driven methods can be used to optimize the performance of industrial SDSs. Naturally, the optimization of speech recognition and understanding performance is one of the most important objectives of an industrial SDS where every retry costs money and increases the likelihood of the call being escalated to a human agent. Using the wealth of data collected once a system is deployed in conjunction with a rigorous continuous transcription/annotation/quality assurance/SLM tuning cycle is a major source of improvement, but not the only one. We have also seen that, depending on the business model and parameters adopted (e.g., costs and gains), one can maximize the return and minimize the caller's frustration by predicting which calls are most likely not to be automated and transferring them to a human agent at earliest convenience. We showed that one can create a classifier able to effectively predict call success at an early moment in a call whose escalation decision improved the overall performance of the SDS in terms of typical industrial business models. A simpler but even more effective method is based on pruning branches of the call flow leading to long but most likely nonautomated calls. Another method of shortening the average handling time of calls and thereby increasing application performance is based on information-theoretical measures to optimize the sequence of questions asked to a caller with the goal of identifying certain pieces of information. We also demonstrated how the dialog strategy of industrial SDSs can be optimized with a method called *Contender* that uses exploration and exploitation to allow for automatically choosing the optimum among competing designs. This tool converts industrial VUI design, historically a manual task, into a semi-automated process. Considering that reliability is a top priority in industrial applications taking huge numbers of complex calls, Contenders may process hundreds of millions of call logs requiring comprehensive management tools. Contenders do not only support designers in building applications best matching the real-world requirements but may substantially increase the applications' performance. Finally, we discussed the recent emergence of cloud-based spoken language processing technology in the smartphone landscape. Main trends include voice search and the spoken dialog agent Siri whose main difference is the underlying search paradigm (textual vs. semantic) as well as the type of result presented to the user (links vs. answers). We do not know whether these two trends will continue to coexist and evolve independently or whether they will merge into a universal, intelligent search agent with deep understanding of the semantics behind text and media on the Web.

References

1. Acomb, K., Bloom, J., Dayanidhi, K., Hunter, P., Krogh, P., Levin, E. Pieraccini, R.: Technical Support Dialog Systems: Issues, Problems, and Solutions. In: Proc. of the HLT-NAACL, Rochester, USA (2007)
2. Cohen, M., Giangola, J., Balogh, J.: *Voice User Interface Design*. Addison-Wesley, Redwood City, USA (2004)
3. Cohen, W., Singer, Y.: A Simple, Fast, and Effective Rule Learner. In: Proc. of the National Conference on Artificial Intelligence, Orlando, USA (1999)
4. Eckert, W., Levin, E., Pieraccini, R.: User modelling for spoken dialogue system evaluation. In: Proceedings of ASRU, pp. 80–87 (1997)
5. Gasic, M., Jurcicek, F., Keizer, S., Mairesse, F., Thomson, B., Yu, K., Young, S.: Gaussian processes for fast policy optimisation of a pomdp dialogue manager for a real-world task. In: Proceedings of SIGDIAL, 2010
6. Gorin, A.L., Riccardi, G., Wright, J.H.: How may I help you? Speech Commun. **23**(1/2), 113–127 (1997)
7. Hunt, A., McGlashan, S.: Speech Recognition Grammar Specification Version 1.0. W3C Recommendation. `http://www.w3.org/TR/2004/REC-speech-grammar-20040316` (2004)

8. Levin, E., Pieraccini, R., Eckert, W.: A stochastic model of human-machine interaction for learning dialog strategies. IEEE Trans. Speech. Audio. Process. **8**(1), 11–23 (2000)
9. Levin, E., Pieraccini, R.: Value-Based Optimal Decision for Dialog Systems. In: Proc. of the SLT, Palm Beach, Aruba (2006)
10. Mitchell, T.: Machine Learning. McGraw Hill, New York, USA (1997)
11. Pieraccini, R., Lubensky, D.: Spoken language communication with machines: The long and winding road from research to business. In: Ali, M., Esposito, F. (eds.), Innovations in Applied Artificial Intelligence. Springer, New York, USA (2005)
12. Pietquin, O., Geist, M., Chandramohan, S., Frezza-Buet, H.: Sample-Efficient Batch Reinforcement Learning for Dialogue Management Optimization. ACM Transactions on Speech and Language Processing **7**(3), 21 (2011)
13. Suendermann, D., Liscombe, V., Evanini, K., Dayanidhi, K., Pieraccini, R., From Rule-Based to Statistical Grammars: Continuous Improvement of Large-Scale Spoken Dialog Systems. In: Proc. of the ICASSP, Taipei, Taiwan (2009)
14. Suendermann, D., Liscombe, J., Pieraccini, R.: Minimally Invasive Surgery for Spoken Dialog Systems. In: Proc. of the Interspeech, Makuhari, Japan (2010)
15. Suendermann, D., Liscombe, J., Pieraccini, R.: Contender. In: Proc. of the SLT, Berkeley, USA (2010)
16. Suendermann, D., Liscombe, J., Pieraccini, R.: Optimize the Obvious: Automatic Call Flow Generation. In: Proc. of the ICASSP, Dallas, USA (2010)
17. Suendermann, D., Liscombe, J., Bloom, J., Li, G., Pieraccini, R.: Large-Scale Experiments on Data-Driven Design of Commercial Spoken Dialog Systems. In: Proc. of the Interspeech, Florence, Italy (2011)
18. Williams, J., Poupart, P., Young, S.: Partially observable markov decision processes with continuous observations for dialogue management. In: Proc. SigDial'06 (2006)

Chapter 9
Conclusion and Future Research Directions

Olivier Pietquin

Processing of speech and language inputs has been studied under a statistical point of view for quite some time. Data-driven methods pioneered by speech recognition researchers such as Rabiner [8] and Jelinek [2] in the late 1970s were applied to natural language understanding only in the early 1990s [5]. It is only a decade ago that dialogue management has benefited from statistical modeling and data-driven methods [3]. Following this trend, this book described the recent advances in statistical data-driven methods for spoken dialogue systems, especially within the European CLASSiC project funded under the seventh framework program. The aim of this project, as reflected by this book, was to produce generic methods for statistical optimization from end to end of a spoken dialogue system, starting with speech recognition and ending with speech synthesis. Machine learning techniques, such as reinforcement learning, were expected to provide useful approaches to this problem because of their ability to solve sequential decision-making problems but also because they rely on a strong mathematical background and on interpretable optimization criteria. Reinforcement learning has thus been applied for optimizing dialogue management and natural language generation (NLG) but also, to some extent, to produce user simulation techniques. Other machine learning methods, such as support vector machines (SVM) or Bayesian networks, were also integrated to make a fully operational system. This book summarized the technical and practical results obtained during this project.

To start with, Chap. 2 introduced statistical approaches to dialogue management, described available data sets, and addressed the problem of data collection. This topic is of primary importance for the success of data-driven methods, one major problem being the collection and, above all, the annotation of data. This chapter proposed a modified approach of the Wizard of Oz paradigm which offers an elegant solution to the data-collection problem and allowed bootstrapping the

O. Pietquin (✉)
SUPELEC - UMI 2958 (GeorgiaTech - CNRS), 2 rue Edouard Belin, 57070 Metz, France
e-mail: olivier.pietquin@supelec.fr

O. Lemon and O. Pietquin (eds.), *Data-Driven Methods for Adaptive Spoken Dialogue Systems*, DOI 10.1007/978-1-4614-4803-7_9,

system design. However, some problems remain, and, in the future, transferability of learned models will become central with the growing interest for speech-based interfaces. As for speech recognition and synthesis, the properties that data corpora should demonstrate to enable transfer of models from task to task, from as little data as possible, will have to be studied further.

Even though spoken language understanding (SLU) has benefited from statistical modeling for more than twenty years now, research in this field is still very active because spoken language is hardly comparable to written language. Spoken language generally contains speech repairs, elliptical utterances, and hesitations. It is also missing grammatical consistency. Chapter 3 dealt with problems inherent to this difference, that is, how to process ill-formed utterances and extract meanings from these. Also, the problem of data sparsity was addressed, and methods to generalize to unseen situations were proposed like the semantic tuple classifier (STC) relying not only on SVM, but also the use of general-domain semantic role labeling. These methods were demonstrated to outperform standard approaches and were successfully applied to an SDS in the TownInfo domain while trained on artificially generated corpora.

Artificially generated data has been used since the early stages of research in the domain of SDS dialogue management optimization [3]. Indeed, popular reinforcement learning algorithms in use in the late 1990s were particularly sample inefficient. This led to a trend of research about user simulation, the problem being to reproduce artificially interactions between a human user and the dialogue manager to be trained. Chapter 4 focused on this problem and presented three different approaches to user simulation, all based on a statistical model of user behavior and trained on data: the agenda-based approach, the dynamic Bayesian network approach and the inverse reinforcement learning one. All these methods aim at generating sequences of decisions that are correlated with data but especially aim at dealing with the lack of annotations. Indeed, user goals and real intentions are most often hidden to the annotator. Anyway, the future of these techniques is unclear since efficient RL algorithms have recently been applied to SDS optimization allowing direct learning from batches of data [4, 6] or to learn online with real users in a reasonable amount of interactions [1, 7]. However, these methods are not really compatible with the ability of human users to adapt their behavior to changes in the SDS management policy. Therefore, user simulation will probably have to be studied in a different way in the future but is likely to remain as an important research area.

Since speech input processing methods are still imperfect, users' utterance meanings are difficult to extract from the speech signal. The context has to be tracked over turns which generally results in several hypotheses based on the history of the interaction. As a consequence, the standard Markov decision process (MDP) paradigm, used to originally optimize SDS management strategies, has been supplemented by the partially observable MDP paradigm. Chapter 5 focused on this paradigm and described the recent advances in the field of optimisation of POMDP-based dialogue managers. Indeed, taking into account partial observability in RL increases greatly the complexity of algorithms in general. Yet, the particular application to SDS can be made tractable thanks to an appropriate modeling of the task. This modeling should not be too much dedicated to the SDS task itself to reduce the impact on the genericity of the data-driven approach. This is why methods to learn model parameters from data were presented in addition to sample-efficient methods for learning interaction policies, based on Gaussian processes and reinforcement learning. In any case, learning generic representation of dialogue states and the tracking of the context based on these representations are still hard to obtain. This area will probably be of major interest in the future years.

The acceptability of speech-enabled interfaces mainly relies on the way information is presented to users. If the machine provides unexpected feedback in terms of content or speech synthesis, it is likely that users will not consider the machine as reliable. Also, the ability of the machine to adapt its behavior according to the user (adaptivity) is often practically implemented in the expressions used to provide information. Chapter 6 addressed the problem of data-driven NLG methods to render systems' outputs adapted to users. NLG is seen as a planning under uncertainty problem and can thus be solved by means of reinforcement learning. This approach conducted the optimisation of adaptive information presentation strategies, resulting in an optimal way of providing information to the users so as to maximize task success (e.g., by optimising the cognitive load induced by the information presentation). Also, optimal use of referred expressions, taking into account the level of expertise of users in the task domain, was shown to improve the overall user satisfaction and system efficiency.

The design cycle of a SDS is often made of engineering tasks, public releases, and redesign phases. Of course, data driven methods aim at reducing the number of cycles by improving the results of engineering tasks based on data collected after public releases or prior to any design task. However, improvements have to be assessed numerically. It is actually a tricky task to define what is a real improvement in the domain of man–machine interactions. In Chap. 7, several objective and subjective evaluation techniques were proposed. Recent proposals coming from both the academic and the industrial worlds were explored taking into account the point of view of different actors: the developer, the business operator, and the user. From this chapter, it is clear that there is still room for improvement in finding the best metric to assess SDS performance, mainly because machine learning methods are mostly based on a single optimisation criterion. Multicriteria optimization may thus be a future trend of research.

This book is largely devoted to academic research conducted during the CLASSiC project. However, this project also aimed at producing results transferable to industrial partners and to the general public. For this reason, Chap. 8 shows the benefit brought by statistical methods in industrial systems. In particular, statistical language models, evolving all through the life of the system, are shown to improve drastically the success rate of dialogue systems and reduce the number of transfers to human operators. Most of the calls can be automated, and only expert operators are required which results in a financial gain for the business operator. Also, dialogue strategies are optimized with methods very close to reinforcement learning. It is clear that data-driven methods are more and more present in industrial systems, while this was not the case only a few years ago. Experiments "in the wild" are now possible because of the tremendous amount of calls processed by deployed systems and logged. The gap between academic research and industrial applications in the field of speech-enabled interfaces is thus getting thinner every year.

A lot of improvement has still to be made to produce natural interactions and user-friendly interfaces. Several avenues of research are foreseen. First, the latency in the responses provided by SDS is still too important and highlights the artificialness of the interfaces. "Incremental" dialogue systems, able to produce answers rapidly (before the end of the users' utterances) and that can be interrupted at anytime, are already under study. Incremental spoken dialogue applications running on mobile devices, usable in several different languages, are the topic of the PARLANCE project,[1] following the CLASSiC project. Naturalness of interfaces also means affective interfaces, able to share and produce emotions. This is the topic of the ILHAIRE project[2] funded under the future and emerging technologies (FET) program of the European commission, in which laughing machines are under consideration. Finally, if humans and machines have to eventually share the same space and if they have to collaborate, then man–machine dialogue should take place in the physical world. Situated interaction should be considered in the future, that is, the integration of the physical context into the description of the dialogue state. Real cooperation should occur between men and machines from which a common ground about the task will emerge. In this vision, spoken dialogue systems should probably move to man–robot interfaces with appropriate social interaction skills, as explored by the EC FP7 project JAMES.[3]

References

1. Gasic, M., Jurcicek, F., Thomson, B., Yu, K., Young, S.: On-line policy optimisation of spoken dialogue systems via live interaction with human subjects". In: Proc. of ASRU 2011, Hawaii (USA) (2011)
2. Jelinek, F.: Continuous speech recognition by statisical methods. IEEE Proceedings **64**(4), 532–536 (1976)
3. Levin, E., Pieraccini, R., Eckert, W.: A stochastic model of human-machine interaction for learning dialog strategies. IEEE Trans. Speech. Audio. Process. **8**(1), 11–23 (2000)
4. Li, L., Balakrishnan, S., Williams, J.: Reinforcement Learning for Dialog Management using Least-Squares Policy Iteration and Fast Feature Selection. In: Proc. of InterSpeech'09, Brighton (UK) (2009)
5. Pieraccini, R., Levin, E.: Stochastic representation of semantic structure for speech understanding. Speech Commun. **11**(2-3), 238–288 (1992)

---

[1] www.parlance-project.eu.

[2] www.ilhaire.eu.

[3] www.james-project.eu.

6. Pietquin, O., Geist, M., Chandramohan, S., Frezza-Buet, H.: Sample-Efficient Batch Reinforcement Learning for Dialogue Management Optimization. ACM Trans. Speech. Lang. Process. **7**(3), 7:1–7:21 (2011)
7. Pietquin, O., Geist, M., Chandramohan, S.: Sample Efficient On-line Learning of Optimal Dialogue Policies with Kalman Temporal Differences. In: Proc. of IJCAI 2011, Barcelona, Spain (2011)
8. Rabiner, L.R., Levinson, S.E., Rosenberg, A.E., Wilpon, J.G.: Speaker independent recognition of isolated words using clustering techniques. IEEE Transactions on Acoustics, Speech and Signal Processing AASP- **27**(4), 336–349 (1979)

MIX
Papier aus verantwortungsvollen Quellen
Paper from responsible sources
FSC® C105338

If you have any concerns about our products,
you can contact us on
ProductSafety@springernature.com

In case Publisher is established outside the EU,
the EU authorized representative is:
Springer Nature Customer Service Center GmbH
Europaplatz 3, 69115 Heidelberg, Germany

Printed by Libri Plureos GmbH
in Hamburg, Germany